GUIDELINES FOR DEVELOPING QUANTITATIVE SAFETY RISK CRITERIA

GUIDELINES FOR DEVELOPING QUANTITATIVE SAFETY RISK CRITERIA

Center for Chemical Process Safety
New York, New York

CENTER FOR
CHEMICAL PROCESS SAFETY
An AIChE Industry
Technology Alliance

WILEY

A JOHN WILEY & SONS, INC., PUBLICATION

It is sincerely hoped that the information presented in this document will lead to an even more impressive safety record for the entire industry. However, neither the American Institute of Chemical Engineers, its consultants, CCPS Technical Steering Committee and Subcommittee members, their employers, their employers' officers and directors, nor ABS Consulting Inc. and its employees warrant or represent, expressly or by implication, the correctness or accuracy of the content of the information presented in this document. As between (1) American Institute of Chemical Engineers, its consultants, CCPS Technical Steering Committee and Subcommittee members, their employers, their employers' officers and directors, and ABS Consulting Inc. and its employees and (2) the user of this document, the user accepts any legal liability or responsibility whatsoever for the consequence of its use or misuse.

A Joint Publication of the Center for Chemical Process Safety of the American Institute of Chemical Engineers and John Wiley & Sons, Inc.

Published by John Wiley & Sons, Inc., Hoboken, New Jersey.
Published simultaneously in Canada.

No part of this publication may be reproduced, stored in a retrieval system, or transmitted in any form or by any means, electronic, mechanical, photocopying, recording, scanning, or otherwise, except as permitted under Section 107 or 108 of the 1976 United States Copyright Act, without either the prior written permission of the Publisher, or authorization through payment of the appropriate per-copy fee to the Copyright Clearance Center, Inc., 222 Rosewood Drive, Danvers, MA 01923, (978) 750-8400, fax (978) 750-4470, or on the web at www.copyright.com. Requests to the Publisher for permission should be addressed to the Permissions Department, John Wiley & Sons, Inc., 111 River Street, Hoboken, NJ 07030, (201) 748-6011, fax (201) 748-6008, or online at http://www.wiley.com/go/permission.

Limit of Liability/Disclaimer of Warranty: While the publisher and author have used their best efforts in preparing this book, they make no representations or warranties with respect to the accuracy or completeness of the contents of this book and specifically disclaim any implied warranties of merchantability or fitness for a particular purpose. No warranty may be created or extended by sales representatives or written sales materials. The advice and strategies contained herein may not be suitable for your situation. You should consult with a professional where appropriate. Neither the publisher nor author shall be liable for any loss of profit or any other commercial damages, including but not limited to special, incidental, consequential, or other damages.

For general information on our other products and services or for technical support, please contact our Customer Care Department within the United States at (800) 762-2974, outside the United States at (317) 572-3993 or fax (317) 572-4002.

Wiley also publishes its books in a variety of electronic formats. Some content that appears in print may not be available in electronic format. For information about Wiley products, visit our web site at www.wiley.com.

Library of Congress Cataloging-in-Publication Data is available.

Guidelines for developing quantitative safety risk criteria / Center for Chemical Process Safety.
    p. cm.
Includes index.
ISBN 978-0-470-26140-8 (cloth)
1. Chemical plants—Accidents--Risk assessment. 2. Chemical industry—Accidents—Prevention.
I. American Institute of Chemical Engineers. Center for Chemical Process Safety.
    TP150.A23G85 2009
    660'.2804—dc22   2009019922

Printed in the United States of America.

10 9 8 7 6 5 4 3 2 1

CONTENTS

PREFACE

The American Institute of Chemical Engineers (AIChE) has been involved with process safety and loss prevention in the chemical, petrochemical, hydrocarbon processing, and other process industries since its founding in 1908. As these industries increased in scale and economic importance in the years following World War II, AIChE's activities in this area expanded through specific programming activities and publications. AIChE's symposia, journals, and other publications have been important safety resources for chemical engineers, industrial chemists, and other technical professionals in these industries.

In 1985, following major chemical disasters in Mexico City, Mexico, and Bhopal, India, the Center for Chemical Process Safety (CCPS) was established as an AIChE Technology Alliance. The charter of CCPS is to develop and disseminate technical information for use in the prevention of major chemical accidents. CCPS is supported by more than 120 sponsors worldwide, including process industry companies, engineering consulting companies, and government entities. These sponsors provide funding, and, more importantly, technical knowledge, expertise, and professional guidance through participation in various CCPS project subcommittees. One important product of CCPS, in support of its charter, is a series of guidelines books which provide information on implementation of various aspects of a complete process safety management system.

Process hazard analysis, risk analysis, and risk management have been recognized as critical features of a process safety management program by CCPS from its inception in 1985. The first product of CCPS was the first edition of *Guidelines for Hazard Evaluation Procedures*, published in 1985, and now in its third edition (published in 2008). The second and third editions of this book include discussion of qualitative risk evaluation tools, which can be incorporated into process hazard studies. CCPS has also published several important books related to process risk analysis, including:

- *Guidelines for Chemical Process Quantitative Risk Analysis* (1989, 2nd Edition published in 2000) provides a complete methodology for quantitative process risk analysis.
- *Evaluating Process Safety in the Chemical Industry: A User's Guide to Quantitative Risk Analysis* (2000) is an overview of the use of quantitative risk analysis techniques in the process industries for engineers and managers.
- *Tools for Making Acute Risk Decisions with Chemical Process Safety Applications* (1995) describes how various decision analysis techniques can be applied to management decision making incorporating the results of risk analysis studies.
- *Layer of Protection Analysis: Simplified Process Risk Assessment* (2001) provides a description of a simplified technique for evaluating process safety risks.

This book, *Guidelines for Developing Quantitative Safety Risk Criteria*, provides additional guidance on the use of risk analysis studies for process risk management decisions. Over the last several decades, as quantitative risk analysis has been applied increasingly in the process industries, many government and industry organizations have begun to develop and apply risk criteria as tools for making process risk management decisions. This book surveys experience with the development and application of these criteria, and provides guidance on how an organization might use this experience to establish its own risk criteria. The book does not recommend criteria, or endorse specific criteria that have been used by various organizations. The focus is on understanding the meaning of the various criteria that have been used, their bases, how they are applied, and how this experience can be used by an organization that is considering similar approaches as a part of its own process risk management system. We hope that this book will help engineers and managers in the process industries better understand risk criteria, how they can be applied, how they can become one element of a comprehensive process risk management system, as well as understand the limitations and potential problems that might be associated with the use of risk criteria.

ACKNOWLEDGMENTS

CCPS wishes to thank the Risk Criteria Subcommittee for their efforts to prepare *Guidelines for Developing Quantitative Safety Risk Criteria*:

Wayne Chastain	*Eastman Chemical Company*
Don Connolley	*BP*
Mike Considine	*BP*
Les Cunningham	*Merck*
Tom Dileo	*Albemarle*
Ken Harrington	*Chevron Phillips Chemical*
Dennis Hendershot	*CCPS Staff Consultant*
Dave Jones	*Chevron*
Shakeel Kadri	*Air Products and Chemicals, Inc.*
Greg Keeports	*Rohm and Haas Company, retired*
Kevin Klein	*Solutia, Celanese*
Sam Mannan	*Texas A&M University*
Bill Marshall	*Eli Lilly and Company*
Jack McCavit	*CCPS Emeritus*
Chee Ong	*Shell*
Scott Ostrowski	*ExxonMobil*
Tim Overton	*Dow*
Bayan M. Saab	*LyondellBasell Industries*
Mike Schmidt	*Emerson Process Management*
Angela Summers	*SIS-TECH*
Karen Tancredi	*DuPont*
John Wincek	*Croda, Inc.*
Brent Woodland	*Conoco/Phillips*
Mike Yurconic	*Intel*

Walt Frank of Frank Risk Solutions, Inc. (formerly of ABS Consulting) and John Farquharson of ABS Consulting were the authors of the book. Steve Arendt, Myron Casada, and Henrique Paula of ABS Consulting provided valuable contributions to its outline and content. Dr. Paula's past extensive research to identify relevant resources on the development of safety risk criteria throughout the world was of great value in developing the book. Karen Taylor edited the manuscript, Susan Hagemeyer prepared the manuscript for the printer, and Paul Olsen provided graphics support.

CCPS also thanks the following peer reviewers for taking the time to review the final manuscript of the book, and providing extensive and valuable input and comments:

Ben Ale	*Technical University Delft*
Lee Alford	*European Process Safety Centre*
Nasir Ali	*Conoco/Phillips*
Arnold Allemang	*Dow, retired*
Marco Amaral	*ABS Consulting*
Wim Blom	*Chevron*
Sergio D. M. Castro	*ABS Group Services do Brasil, Ltda.*
Mohan Chahal	*ConocoPhillips*
Phil Crossthwaite	*DNV*
Brian Dunbobbin	*Air Products and Chemicals, Inc.*
Leopoldine Montiel Frioni	*CETESB*
Martin Goose	*IChemE Safety & Loss Prevention Subject Group*
Richard C. Griffin	*Chevron Phillips Chemical*
Koos Ham	*TNO Built Environment and Geosciences*
Randy Hawkins	*RRS Schirmer*
Jean-Paul Lacoursiere	*University of Sherbrooke*
Steve Leigh	*Croda International plc*
John R. Lockwood	*Independent risk consultant*
Donald K. Lorenzo	*ABS Consulting*
John Marshall	*Dow, retired*
Georges A. Melhem	*ioMosaic Corporation*
José Luiz Mendes	*INEA*
Thia Cheong Meng	*International Refinery Services Singapore*
Lisa Morrison	*BP International Limited*
Art Mundt	*Dow, retired*
Lou Nathanson	*ConocoPhillips*
Robert Ormsby	*CCPS*
Bill Ostrowski	*ConocoPhillips*
Robin Pitblado	*DNV*
Ana Flávia Sampaio	*ABS Consulting*

Gregg Sapp	*Conoco/Phillips*
Marcos Tadeu Seriacopi	*CETESB*
Irfan Shaikh	*Lloyd's Register North America, Inc.*
Diogo Singulani	*INEA*
Emery Thomas	*ConocoPhillips*
Hal Thomas	*Air Products and Chemicals, Inc.*

LIST OF TABLES

LIST OF FIGURES

ACRONYMS AND ABBREVIATIONS

ACDS	Advisory Committee on Dangerous Substances (UK)
ACMH	Advisory Committee on Major Hazards (UK)
AIChE	American Institute of Chemical Engineers
ALARA	As Low As Reasonably Achievable
ALARP	As Low As Reasonably Practicable
API	American Petroleum Institute
AQPP	Air Quality Permitting Program (New Jersey)
BLS	Bureau of Labor Statistics (US)
BRZO	Besluit Risico's Zware Ongevallen (Major Incident Risk Order, Netherlands)
CCPHI	Coordinating Committee on Land-Use Planning and Control Relating to Potentially Hazardous Installations (Hong Kong)
CCPS	Center for Chemical Process Safety
CETESB	Companhia de Tecnologia de Saneamento Ambiental (Brazil)
CHEM Services	Chemical Hazards and Emergency Management Services (Queensland, Australia)
COMAH	Control of Major Accident Hazards regulation (UK)
COVO	Commissie Veiligheid Omwonenden (commission for safety of inhabitants) of the Rijnmond Authority (Netherlands)
CPQRA	chemical process quantitative risk analysis
CSChE	Canadian Society for Chemical Engineering
DOD	Department of Defense (US)
DOE	Department of Energy (US)

DOP	Department of Planning (New South Wales, Australia)
EIA	environmental impact assessment
EIAO	Environmental Impact Assessment Ordinance (Hong Kong)
EPA	Environmental Protection Agency (US)
EPD	Environmental Protection Department (Hong Kong)
FAR	fatal accident rate
FDA	Food and Drug Administration (US)
FEEMA	Fundação Estadual de Engenharia do Meio Ambiente (Brazil)
FEPAM	Fundação Estadual de Proteção Ambiental (Brazil)
FMEA	failure modes and effects analysis
F-N	frequency-number (curve)
FSA	formal safety assessment
HAZID	hazard identification
HAZOP	hazard and operability analysis
HID	Hazardous Installations Directorate (UK)
HSE	Health and Safety Executive (UK)
IBAMA	Brazilian Institute of Environment and Renewable Natural Resources
IDLH	immediately dangerous to life or health
IMO	International Maritime Organization
IPEEE	Individual Plant Examination of External Events
IPL	independent protection layer
IPS	instrumented protective system
LFL	lower flammable limit
LNG	liquefied natural gas
LPG	liquefied petroleum gas
LOPA	layer of protection analysis
MHF	Major Hazard Facilities Regulation (Victoria, Australia)
MHIDAS	Major Hazard Incident Data Service database (EU)
MI	mechanical integrity
MIACC	Major Industrial Accidents Council of Canada
NASA	National Aeronautics and Space Administration (US)
NJDEP	New Jersey Department of Environmental Protection
NRC	Nuclear Regulatory Commission (US)

NSC	National Safety Council (US)
NSW	New South Wales, Australia
OSHA	Occupational Safety and Health Administration (US)
PADHI	Planning Advice for Developments Near Hazardous Installations (UK)
PCD	Pollution Control Department (Singapore)
PHA	process hazard analysis
PHI	potentially hazardous installation
PLL	potential loss of life
PPRT	Technological Risks Prevention Plan (France)
PRA	probabilistic risk assessment
QRA	quantitative risk assessment
RBPS	risk-based process safety
RSSG	Royal Society Study Group (UK)
SFAIRP	So Far As Is Reasonably Practicable
SIS	safety instrumented system
TOR	tolerability of risk
TQ	threshold quantity
V&W	Ministerie van Verkeer en Waterstaat (Ministry of Transport, Public Works, and Water Management, Netherlands)
VCE	vapor cloud explosion
VROM	Volkshuisvesting, Ruimtelijke Ordening en Milieu (Ministry of Housing, Spatial Planning, and the Environment, Netherlands)

GLOSSARY

Aggregate risk: Societal risk for on-site workers in occupied buildings (uniquely applied to facility siting studies, as described in [CCPS 1996] and [API 2003]).

Anchor point: A single F-N pair which, along with the slope, defines an F-N risk criterion curve.

Apportionment: The subdividing of a risk criterion among a number of risk sources (for example, among all process units at an operating site). The concept of risk apportionment can apply to both individual and societal risk, but at different levels in the enterprise.

As Low As Reasonably Achievable (ALARA): A concept equivalent to *As Low As Reasonably Practicable*.

As Low As Reasonably Practicable (ALARP): The concept that efforts to reduce risk should be continued until the incremental sacrifice (in terms of time, effort, cost, or other expenditure of resources) is grossly disproportionate to the value of the incremental risk reduction achieved.

Average individual risk (exposed hours/worked hours): The individual risk for an activity, calculated for the duration of the activity, or averaged over the working day.

Average individual risk (exposed population): The individual risk averaged over the population which is exposed to risk from the facility.

Average individual risk (total population): The individual risk averaged over a predetermined population, without regard to whether or not all people in that population are actually exposed to the risk.

Average rate of death: A measure of societal risk expressed as the average number of fatalities that might be expected per unit time from all possible incidents.

Canvey Island Study: An early (1978) quantitative risk assessment which evaluated the risks associated with an integrated chemical/petrochemical processing complex in the UK.

Checklist analysis: A hazard evaluation procedure using one or more pre-prepared lists of process safety considerations to prompt team discussions of whether the existing safeguards are adequate.

Consequence: The undesirable result of an incident, usually measured in health and safety effects, environmental impacts, loss of property, and business interruption costs.

Consequence modeling: Development of numerical estimates of the expected effects of incident outcome cases independent of frequency or probability.

De minimis risk: A level of risk that would be perceived by most to be broadly acceptable, and not requiring further reduction.

Establishment: In the context of the EU Seveso Directive (Seveso 2), the whole area under the control of an operator where dangerous substances are present in one or more installations, including common or related infrastructures or activities.

Event tree: A logic model that graphically portrays the combinations of events and circumstances in an accident sequence.

Event tree analysis: A method used for modeling the propagation of an initiating event through the sequence of possible incident outcomes. The event is represented graphically by a tree with branches from the initiating cause through the success or failure of independent protection layers.

Experience-based criteria: Risk criteria that are calibrated/validated against established consensus as reflected in historical precedents/decisions about tolerance of major hazard risks, particularly those subjected to broad public discussion or scrutiny.

External safety: A term used in the Netherlands to refer to "risks to the surrounding area in handling hazardous materials, in the use, storage, and transport of those materials." [VROM 2005]

Failure modes and effects analysis (FMEA): A hazard evaluation procedure in which all known failure modes of components or features of a system are considered in turn and undesired outcomes are noted to guide discussions of whether the existing safeguards are adequate.

Fatal accident rate: A measure of individual risk expressed as the estimated number of fatalities per 10^8 exposure hours (roughly 1000 employee working lifetimes).

Fault tree analysis: A method used to analyze graphically the failure logic of a given event, to identify various failure scenarios (called cut-sets), and to support the probabilistic estimation of the frequency of the event.

F-N curve: A plot of cumulative frequency versus consequences (typically expressed as number of fatalities).

Frequency modeling: Development of numerical estimates of the likelihood of an event occurring.

Hazard and operability analysis (HAZOP): A scenario-based hazard evaluation procedure in which a team uses a series of guide words to identify possible deviations from the intended design or operation of a process, then examines the potential consequences of the deviations and the adequacy of existing safeguards.

HSE Risk Interval: A single-value measure of societal risk that can be calculated from the data used to develop an *F-N curve*, or can be estimated based upon an assumed shape for the F-N curve and information on a proposed "worst case" scenario.

Hypothetical individual: A modeling construct that allows the risk estimate to reflect variations in exposure durations or patterns, diverse forms of protection available to different population groups, variations in individual characteristics that might affect susceptibility to harm (e.g., age, gender, abilities or disabilities), and so forth. Multiple risk calculations may be required to determine the hypothetical individual at greatest risk at a particular location.

Immediately dangerous to life or health: An atmospheric concentration of any toxic, corrosive, or asphyxiant substance that poses an immediate threat to life or would cause irreversible or delayed adverse health effects or would interfere with an individual's ability to escape from a dangerous atmosphere.

Incident: An event, or series of events, resulting in one or more undesirable consequences, such as harm to people, damage to the environment, or asset/business losses. Such events include fires, explosions, releases of toxic or otherwise harmful substances, and so forth.

Independent protection layer (IPL): A device, system, or action that is capable of preventing a postulated accident sequence from proceeding to a defined, undesirable endpoint. An IPL is (1) independent of the event that initiated the accident sequence and (2) independent of any other IPLs. IPLs are normally identified during layer of protection analyses.

Individual risk: The risk to a person in the vicinity of a hazard. This includes the nature of the injury to the individual, the likelihood of the injury occurring, and the time period over which the injury might occur.

Installation: In the context of the EU Seveso Directive (Seveso 2), a technical unit within an establishment in which dangerous substances are produced, used, handled, or stored. It includes all the equipment, structures, pipework, machinery, tools, private railway sidings, docks, unloading quays serving the installation, jetties, warehouses or similar structures, floating or otherwise, necessary for the operation of the installation.

Instrumented protective system (IPS): A safety system composed of a separate and independent combination of sensors, logic solvers, final elements, and support systems that are designed and managed to achieve a specified risk reduction.

Involuntary risk: Risk that is imposed on someone who does not directly benefit from the activity that poses the risk.

Land use planning: The control of developments around hazardous installations based upon consideration of the risks posed by the installations, and the nature of the developments and the populations they might contain.

Layer of protection analysis (LOPA): An approach that analyzes one incident scenario (cause/consequence pair) at a time, using predefined values for the initiating cause frequency, independent protection layer failure probabilities, and consequence severity, in order to compare an order-of-magnitude scenario risk estimate to tolerable risk goals for determining where additional risk reduction or more detailed analysis is needed. Scenarios are identified elsewhere, typically using a scenario-based hazard evaluation procedure such as a HAZOP study.

Less vulnerable object: In the Dutch regulatory context, an off-site population characterized by a lower population density of generally more fit and/or more easily evacuated persons, as contrasted with a *vulnerable object*.

Likelihood: A measure of the expected frequency with which an event occurs. This may be expressed as a frequency (e.g., events per year), a probability of occurrence during a time interval (e.g., annual probability), or a conditional probability (e.g., probability of occurrence, given that a precursor event has occurred).

Location risk: Individual risk calculated for a particular geographical location, independent of the nature of the population, or whether anyone is likely to be present at a particular location. The calculation assumes a receptor present 24 hours per day, 365 days per year, out of doors, with no protection.

Maximum individual risk: The individual risk to the person(s) exposed to the highest risk in an exposed population.

Off-site population: Persons located outside of the site property line that may be impacted by an on-site incident.

Probit: A random variable with a mean of 5 and a variance of 1, which is used in various effect models. Probit-based models derived from experimental dose-response data, are often used to estimate the health effect that might result based upon the intensity and duration of an exposure to a harmful substance or condition (e.g., exposure to a toxic atmosphere, or a thermal radiation exposure).

Process hazard analysis: An organized effort to identify and evaluate hazards associated with processes and operations to enable their control. This review normally involves the use of qualitative techniques to identify and assess the significance of hazards. Conclusions and appropriate recommendations are developed. Occasionally, quantitative methods are used to help prioritized risk reduction.

Quantitative risk assessment (QRA): The systematic development of numerical estimates of the expected frequency and consequence of potential accidents associated with a facility or operation based on engineering evaluation and mathematical techniques.

Rijnmond Study: An early (1982) quantitative risk assessment that evaluated the risks associated with an integrated chemical/petrochemical process complex in the Netherlands.

Risk: A measure of human injury, environmental damage, or economic loss in terms of both the incident likelihood and the magnitude of the loss or injury. A simplified version of this relationship expresses *risk* as the product of the *likelihood* and the *consequences* (i.e., Risk = Consequence x Likelihood) of an incident.

Risk acceptance criteria: See *Risk tolerance criteria.*

Risk analysis: The development of a quantitative estimate of risk based on engineering evaluation and mathematical techniques for combining estimates of incident consequences and frequencies.

Risk assessment: The process by which the results of a risk analysis (i.e., risk estimates) are used to make decisions, either through relative ranking of risk reduction strategies or through comparison with risk targets [risk criteria]. Note, these *Guidelines* use, for simplicity in presentation, the single term *risk assessment* to also include those activities addressed in the definition of *risk analysis.*

Risk-based process safety: The CCPS process safety management system approach that uses risk-based strategies and implementation tactics that are commensurate with the risk-based need for process safety activities, availability of resources, and existing process safety culture to design, correct, and improve process safety management activities.

Risk contour: Lines that connect points of equal risk around the facility ("isorisk" lines).

Risk escalation: A risk management system whereby an increasingly higher level of authorization is required to sanction the continued tolerance of increasingly higher levels of risk.   Some organizations use the term risk elevation.

Risk management: The systematic application of management policies, procedures, and practices to the tasks of analyzing, assessing, and controlling risk in order to protect employees, the general public, and the environment as well as company assets while avoiding business interruptions.

Risk matrix: A tabular approach for presenting risk tolerance criteria, typically involving graduated scales of *incident likelihood* on the Y-axis and *incident consequences* on the X-Axis.   Each cell in the table (at intersecting values of *incident likelihood* and *incident consequences*) represents a particular level of risk.

Risk tolerance criteria: A predetermined measure of risk used to aid decisions about whether further efforts to reduce the risk are warranted.

Risk tolerance: This has been described as "willingness by society as a whole to live with a risk so as to secure certain benefits in the confidence that the risk is one that is worth taking and that it is being properly controlled. However, it does not imply that ... everyone would agree without reservation to take this risk or have it imposed on them." [HSE 2001]

Scaling: Adjusting the relative size of a risk criterion, at a given level in the enterprise, based upon factors such as scale of operation. For example, the societal risk criterion for a business sector might be apportioned to the various sites within the sector with sites receiving differing shares of the allotment. Scaling is not applicable to individual risk.

Scenario: An unplanned event or incident sequence that results in a loss event and its associated impacts, including the success or failure of safeguards involved in the incident sequence.

So Far As Is Reasonably Practicable (SFAIRP): A concept that is generally equivalent to *As Low As Reasonably Practicable* (however, there are some legal distinctions between the two terms in the UK).

Societal risk: A measure of risk to a group of people. It is most often expressed in terms of the frequency distribution of multiple casualty events.

Technology-driving criteria: Risk criteria that represent aggressive goals, whether or not experience indicates that such goals are currently attainable. The implication is that the goals will drive the development of the means for meeting the criteria.

Voluntary risk: Risk that is consciously tolerated by someone seeking to obtain the benefits of the activity that poses the risk.

Vulnerable object: In the Dutch regulatory context, an off-site population that may be more significantly impacted by a risk source, as contrasted with a *less vulnerable object*. This greater susceptibility may be due to a variety of considerations such as the numbers of persons present, the physical characteristics of the population (e.g., the young or infirm), occupancy patterns (e.g., continuously present versus transient), or constraints on the ability to evacuate. Individual risk criteria for *vulnerable objects* are more rigidly enforced than those for *less vulnerable objects*.

What-if analysis: A scenario-based hazard evaluation procedure using a brainstorming approach in which typically a team that includes one or more persons familiar with the subject process asks questions or voices concerns about what could go wrong, what consequences could ensue, and whether the existing safeguards are adequate.

*The best we can do is size up the chances,
calculate the risks involved, estimate
our ability to deal with them, and then
make our plans with confidence.*

Henry Ford (1863 – 1947)

1
INTRODUCTION

1.1   WHAT IS RISK?

Every day, each of us makes myriad decisions, takes an untold number of actions, and avoids many other actions – and associated with virtually every decision, action, and inaction is some possibility of harm to ourselves, those around us, our property, or our surroundings. The likelihood of experiencing this harm – which we call *risk* and will define further below – and companies' interest in setting limits on process-related risk are the focus of this book.

The risks associated with our decisions, actions, and inactions can range from minimal to very high, and we may or may not consciously recognize those risks. For example, an activity may be so commonplace, such as descending a stairway in our home, that our awareness of the risk is, at best, subliminal. In contrast, because we infrequently access the roof of our house, we may very carefully consider the possibility for harm associated with climbing to the top of a long ladder.

Where risk is recognized, most prudent individuals would seek to avoid or manage that risk. Efforts to do so necessarily prompt such questions as, "Have I done enough (or too much)?" or "Am I putting my resources in the right places?" or "How much risk is too much?" The creation of criteria for the process industries to use as decision aids in addressing such questions is the subject of this book.

1.1.1   Risk Basics

Risk is a concept that is broadly understood in a vague way by most people, and specifically understood in different ways by many people. Man's explicit consideration of risk (at least in the historical era) began at least two millennia ago with the advent of early games of chance and the need to consider what might be a prudent wager. Formalized studies of the concept of risk, as it is now viewed, began in the 16th century. At that time, some students of risk were still focused on dice and card games. Others, however, were concerned with a new form of risk-taking – the need for investors to protect themselves against potential losses associated with the transport of goods by ship on increasingly long sea voyages. Such concerns prompted the development of early predecessors of the insurance industry. [Bernstein 1998]

In about 1662, mathematicians in the Port-Royal monastery in Paris first codified the modern concept of risk when they wrote, "Fear of harm ought to be proportional not merely to the gravity of the harm, but also to the probability of the event." [Arnauld 1662] Fundamentally, this definition of risk has not changed in 350 years. Correspondingly, the Center for Chemical Process Safety (CCPS) has defined risk the following way [CCPS 2000a]:

> "Risk is a measure of human injury, environmental damage, or economic loss in terms of both the incident likelihood and the magnitude of the loss or injury."[1]

Dissecting this definition into its component parts aids in the understanding of risk.

This book addresses unexpected events having undesired results. CCPS uses the term *incident* to describe such events, which run the gamut of fires, explosions, releases of toxic or otherwise harmful substances, and so forth.

The *loss, damage, or injuries* resulting from the incident are termed the *consequences* of the event. In discussing risk in a particular context, the nature of the consequences must be clearly defined. For example, the magnitude of human injury might range from a minor injury, to permanent disability, to death. While all injuries are cause for concern, fatality is the degree of injury most frequently selected when quantifying process-related risks to persons, and it is the degree of injury that will be addressed in these *Guidelines*.

The *likelihood* of the incident expresses how frequently the incident is expected to occur or, more precisely, how frequently the defined level of consequences is expected to be experienced. This may be expressed as a frequency (or a probability of occurrence during some time interval).

---

[1]The UK Health and Safety Executive provides an analogous definition of risk [HSE 2007a]: "The likelihood of a specified undesired event occurring within a specified period or in specified circumstances."

Algebraically, *risk* can be represented as:

$$Risk = f(consequence, likelihood)$$

A simplified version of this relationship, which is often appropriate, expresses *risk* as the product of the *likelihood* and the *consequences* (i.e., Risk = Consequence x Likelihood). Thus, if *consequences* are expressed in terms of loss per incident, and *likelihood* as the number of incidents per year, the calculated *risk* will have the units of losses per year.

Equivalent Risks 0.1 Losses per Year	
Losses/Event	Events/Year
1	1/10
10	1/100
100	1/1,000
3,000	1/30,000

This equation shows that a numerically equal value of risk can result from an unlimited number of frequency and consequence pairs (see sidebar). A single loss every 10 years and 100 losses every 1,000 years yield the same value of risk. However, these *Guidelines* will discuss later how risk perceptions can change as the magnitude of the consequences increases.

Several important concepts should be kept in mind with respect to the evaluation and management of risk.

First, no such thing as zero risk exists. While industry may seek to reduce and manage the risks associated with process operations, some (hopefully low) level of residual risk will always remain.

Thus, the only way to eliminate the risk of a particular industrial activity is to cease the activity (e.g., shutting down a particular chemical process or no longer handling a particular chemical). However, such risk-reduction decisions may have unforeseen risk trade-offs (e.g., a decision to stop chlorinating drinking water in the attempt to avoid the hazards of handling chlorine may lead to an increase in water-borne diseases). Staying in bed might be seen as a means of avoiding a variety of hazards and the associated fatality risks (e.g., slipping on a bar of soap in the shower, falling down the stairs, experiencing a car accident while driving to work). However, this course of inaction has its own attendant risks (deteriorating health as a consequence of a sedentary lifestyle, falling out of bed, dying in a house fire while sleeping). Any human activity carries with it the potential for an undesired negative consequence. [Hendershot 1997]

Second, risk management requires resources. While some initial gains may be relatively inexpensive, each incremental reduction in risk will likely require a greater expenditure of resources than did the previous. Many risk analysts assert that the most prudent risk management strategy, therefore, is the one that distributes available resources in a way that effects the greatest overall reduction in the risk of concern to the organization. Decisions of where and when to devote resources, and when not to, are aided by the availability of criteria such as described in these *Guidelines*.

Third, to borrow a phrase from the investment industry, "past performance is no guarantee of future success." To understand risk, we must understand the laws of chance. Flipping a coin and getting heads three times in a row does not mean that heads will be either more or less likely on the fourth toss. Similarly, operating a chemical process for 20 years without a significant process incident does not prove that the process is free of risks. Hopefully, that experience is indicative of a robust safety management system that has successfully managed the risk; however, it is not a vaccination against future harm. Absence of the evidence of risk is not evidence of the absence of risk. [HSE 2001]

Finally, risk tolerance is not the same as risk acceptance. These *Guidelines* address what are often termed *risk tolerance criteria*, not *risk acceptance criteria*. Why make this distinction? No one wants to be harmed; consequently, it is difficult to accept that other individuals or organizations might impose upon us the potential for harm. However, while we may not find this situation to be *acceptable*, a level of risk may exist that we view to be *tolerable* in return for an associated benefit that we seek (see sidebar). Our tolerance may be based, in part, on the confidence that efforts continue to further reduce the risk. [HSE 2001]

For example, air travel exposes passengers to some level of risk. However, that risk is generally tolerated in return for the benefit of being able to get to a destination more quickly. In addition, we may have some confidence that, collectively, federal regulations governing the training of flight crews and the maintenance of aircraft, and the airline's recognition of its own self-interests, stimulate efforts to reduce that risk.

> **The Distinction between**
> ***acceptance* and *tolerance*...**
>
> *Acceptance* - to agree to take (something), or to consider (something) as satisfactory, reasonable, or true
>
> *Tolerance* - the ability to bear something unpleasant or annoying
>
> **[Cambridge 2008]**

1.1.2   Why Is Risk Assessment Important?

While these *Guidelines* focus on establishing *risk criteria*, understanding the role that such criteria play is helpful when considering the broader context of how organizations manage the risk of their activities. The CCPS *Guidelines for Chemical Process Quantitative Risk Analysis* (*CPQRA Guidelines*) defines *risk management* as:

> Risk Management – The systematic application of management policies, procedures, and practices to the tasks of analyzing, assessing, and controlling risk in order to protect employees, the general public, and the environment as well as company assets while avoiding business interruptions. [CCPS 2000a]

The keys to the implementation of a risk management program are the activities of risk analysis and risk assessment, which are defined in the *CPQRA Guidelines* as:

> Risk Analysis – The development of a quantitative estimate of risk based on engineering evaluation and mathematical techniques for combining estimates of incident consequences and frequencies.

> Risk Assessment – The process by which the results of a risk analysis (i.e., risk estimates) are used to make decisions, either through relative ranking of risk reduction strategies or through comparison with risk targets [*risk criteria*].

These activities are discussed in greater detail in the *CPQRA Guidelines*.

While the distinction between risk analysis and risk assessment is important within the context of the *CPQRA Guidelines*, these *Guidelines* will use, for simplicity in presentation, the single term *risk assessment* to aggregate the activities that seek answers to the following questions essential to understanding and managing the risks associated with a chemical process:

- What are the hazards/what can go wrong (scenario)?
- How severe could it be (consequence)?
- How likely is it to happen (frequency)?
- How do consequence and frequency combine (risk)?
- Is the current level of risk tolerable, considering existing safeguards?
- If not, what needs to be done to reduce and manage the risk?

Experienced process safety practitioners will recognize these questions as fundamental in the conduct of a thorough *process hazard analysis* (PHA). Indeed, a qualitative analysis, such as a PHA, is a prerequisite for any evaluation of process risk. However, PHAs and similar analyses often identify many high-consequence/low-probability scenarios that are very difficult to address is a qualitative analysis – their scopes, in terms of either consequence or likelihood, are too far beyond the experience base of the PHA team to be handled qualitatively. In these situations, quantitative risk assessment (QRA) becomes a useful tool for gaining perspective on the collective significance of these potential events.

In a QRA, the estimated consequences and the estimated frequency of each scenario are combined to estimate the associated risk. At that point, the need for risk criteria arises. Having an estimate of the level of risk associated with an activity does not directly manage or reduce that risk. Improvements in safety (i.e., risk reduction), when necessary, require making a decision to change something, followed by action to effect that change. The primary reason to examine risk is to assist in making such decisions.

Risk criteria are useful in examining and judging the significance of risk. Producing risk results without having understandable criteria for judging them would be similar to a teacher assigning grades without providing a way to understand what an A, B, or F means. These *Guidelines* serve to explain such risk criteria and to describe how a company might go about developing them.

Specific benefits from risk assessment as part of a risk management system include:

- Providing a logical way to analyze risks, which should increase confidence that risk management decisions are rationally determined and not arbitrarily made
- Promoting greater consistency in risk-based decision-making across the organization
- Providing a basis for prioritizing/apportioning finite resources (providing the best mix of expenditures to minimize total risk across the company)
- Assisting in the evaluation of the relative benefits of risk-reduction alternatives
- Helping define which level of the organization should take responsibility for the decisions that affect the risk (i.e., higher risk decisions made at higher levels)
- Helping protect the organization's permission to operate (actual or figurative) and enhancing the sustainability of the business
- Yielding a better understanding of the management of the risk

Industry also performs risk assessments because society increasingly expects industry to be cognizant of, and to responsibly manage, the risks of its operations. Indeed, as noted later in this book, the expectations of society will likely play an important and significant role in the determination of the risk criteria applied in risk assessments.

1.1.3   Residual Risk versus "Zero Incidents"

Section 1.1.1 noted that no such thing as "zero risk" exists – that despite diligent efforts, some residual risk is always associated with any human endeavor. Yet, many organizations hold the concept of "zero incidents" as a goal. We could reasonably ask whether the concept of residual risk and the goal of zero incidents are not mutually inconsistent.

Even a company aspiring to a goal of zero incidents will typically have annual targets (finite and non-zero) as part of the path to attaining its zero incident goal. The fact that industry analyzes and assesses risk and, after reasonable risk reduction efforts, finds it must leave some risk on the table should not be construed as denigrating the importance of the goal of zero incidents. Rather, risk assessment should be viewed for what it is, an important tool in a risk management program intended to help bring the organization closer to its goal of zero incidents.

Table 1.1 summarizes CCPS' 20-element risk-based process safety model that is described in its *Guidelines for Risk Based Process Safety (RBPS Guidelines)*. [CCPS 2007a]   Risk assessment, as defined in the current *Guidelines*, is addressed in the element *Hazard Identification and Risk Analysis*. The RBPS model illustrates that risk assessment, while important, is only one of many activities in an integrated risk management system.

1.2   SCOPE OF THESE GUIDELINES

The process industries face a wide variety of risks including, but not limited to, potential harm to workers and plant neighbors as a consequence of the hazardous nature of process materials and activities, associated environmental damage, the resulting damage to the reputation of the company, and economic losses. These *Guidelines* will limit discussion to the risk of human safety impacts from process operations. Specifically:

> **Examples of Other Risks Faced By the Process Industries**
>
> • Damage to infrastructure and equipment
> • Damage to company reputation
> • Danger to industry reputation
> • Environmental damage
> • Business interruption costs
> • Security (terrorism) threats
> • Damage to vital records
> • Chronic health effects

- Consequences associated with episodic events (i.e., events of limited duration – typically an incident such as a fire, explosion, or hazardous material release) and
- Acute health effects (i.e., adverse health effects with severe symptoms developing rapidly) leading to fatalities

TABLE 1.1. Risk Based Process Safety Elements

Commit to Process Safety
- *Process Safety Culture*
- *Compliance with Standards*
- *Process Safety Competency*
- *Workforce Involvement*
- *Stakeholder Outreach*

Understand Hazards and Risk
- *Process Knowledge Management*
- *Hazard Identification and Risk Analysis*

Manage Risk
- *Operating Procedures*
- *Safe Work Practices*
- *Asset Integrity and Reliability*
- *Contractor Management*
- *Training and Performance Assurance*
- *Management of Change*
- *Operational Readiness*
- *Conduct of Operations*
- *Emergency Management*

Learn from Experience
- *Incident Investigation*
- *Measurement and Metrics*
- *Auditing*
- *Management Review and Continuous Improvement*

Thus, for example, these *Guidelines* cover topics that could be applied to addressing the risk of fatal injuries resulting from the toxic effects of chlorine inhaled after a railcar rupture, but do not address the effects of non-life-threatening chlorine burns to the skin.   Similarly, this document does not address issues such as long term cancer deaths from chronic exposures to contaminants in drinking water.

The principles outlined in these *Guidelines* can be applied to many different types of risks, such as those listed in the sidebar on the previous page; however, the discussion of these diverse risks is beyond the scope of this book.

1.3   OBJECTIVES OF THESE GUIDELINES

1.3.1   What these Guidelines Are Intended to Achieve

These *Guidelines* are intended to provide: (1) an appreciation of the importance of risk criteria, (2) a sound understanding of the types and application of risk criteria, (3) good practices to follow in establishing risk criteria, and (4) risk criteria precedents.

These *Guidelines* provide the reader with background information and an awareness of key considerations essential to establishing quantitative risk criteria that are sound, responsible, and workable. Companies that follow the guidance provided in these *Guidelines* should be able to develop risk criteria that are both rational and aligned with societal values. Explicit, well-defined risk criteria, communicated as part of risk assessment activities, add to the clarity of the decision-making process.

Readers should achieve a better understanding of risk and its management, with the goal of providing a basis for:

- Greater consistency in risk decision making across the organization
- Additional insight into the risks that are being managed, as well as the importance of key controls
- More effective communications on risk related issues

Communication of risk information can be difficult – especially if the communications are directed to groups with strong feelings about the issue. Some stakeholders (both internal and external) may feel that they should not have to tolerate any risk – particularly if the bases for risk decisions are not clear or not aligned with societal values. While a detailed discussion of risk communication is beyond the scope of these *Guidelines*, sound risk criteria, developed using the approach set out in these *Guidelines*, should make such communications easier and more productive.

1.3.2   What These Guidelines Do Not Intend

These *Guidelines* do not propose specific risk criteria for the process industries. As will be discussed in detail, the formulation of risk criteria by a company must be accomplished within the context of a variety of considerations (e.g., societal, legal, business, and perhaps, regulatory) that may be unique to the company. The recommendation of specific risk criteria is beyond the charter of CCPS' mission.

Chapter 3 and Appendix B discuss relevant risk criteria that have been promulgated by regulatory agencies and other organizations. The intent is to provide valuable perspective on the topic and to assist in the calibration of company-specific efforts. However, these *Guidelines* do not imply the endorsement of any specific criteria.

The superior man, when resting in safety,
does not forget that danger may come. ...
When all is orderly, he does not forget
that disorder may come.

Confucius (551 BC - 479 BC)

2
FUNDAMENTAL CONCEPTS OF RISK ASSESSMENT AND RISK CRITERIA

Although these *Guidelines* focus on the application of risk criteria in risk assessments, they are not intended to provide guidance on how to conduct a quantitative risk assessment (QRA). Nevertheless, a basic understanding of risk assessment should help the reader better appreciate the basis for, and the application of, risk criteria. This chapter seeks to establish such a foundation, beginning with a concise history of the topic.

2.1   A BRIEF HISTORY OF RISK ASSESSMENT

2.1.1   The Early History

Some of the earliest formalized applications of risk decision making were focused on financial risk – in the context of early insurance initiatives. By about 1700, the government of England was attempting to close its budget

> **A History of Risk**
>
> Readers interested in a historical perspective of risk and its consideration may wish to read *Against the Gods* by Peter L. Bernstein. [Bernstein 1998]

gap by selling life annuities, and early maritime insurance brokers were congregating in Edward Lloyd's coffee house in London. While we may not know the risk criteria they applied, and while some brokers were undoubtedly less than successful, the underlying concepts developed at that time evolved into today's insurance industry.

2.1.2   The Beginnings of the Modern Consideration of Technological Risk

The Dutch Battle with the Sea – The Dutch have centuries of experience in water management and flood control, with the earliest dikes designed to hold back the North Sea being built more than 1,000 years ago. By 1200 AD, the first local water management boards were established in the Netherlands. These boards, elected by the population, were charged with the responsibility for building and maintaining the dikes and, no doubt, faced such questions as, "How high must the dikes be?" or "How much protection is enough?"

A quantitative approach to answering such questions was developed following disastrous floods in 1953, which killed nearly 2,000 people. This natural disaster prompted the government to establish a criterion requiring that an individual's risk of drowning in a susceptible area be less than 1 in a million per year (10^{-6} fatality/year). [Ale 2005a]

Aerospace Beginnings. Pioneering work on QRA began in the early 1960s at Bell Telephone Laboratories to address the need for evaluating missile launch control system reliability. The fault tree technique was created here and later developed further at Boeing. In such early aerospace industry applications, the risk criteria were the reliability goals established for high-consequence, low-frequency events (typically failure of a mission). These goals were used to set reliability targets for systems and components. Fault trees and reliability block diagrams (drawn in success logic

> For a detailed description of fault tree analysis, and other quantitative techniques used in QRA, refer to CCPS' *Guidelines for Chemical Process Quantitative Risk Analysis* (CCPS, 2000a).

rather than the failure logic of fault trees) were used to evaluate the overall system reliability based on individual component and subsystem reliability.

The National Aeronautics and Space Administration (NASA) space shuttle program provides a more recent example of the aerospace application of QRA. Early NASA estimates of the probability of the catastrophic loss of a shuttle were subjective; however, the Challenger disaster prompted more quantitative approaches. Dr. Richard Feynman, a member of the Challenger investigation panel, estimated the probability at 1 in 100 missions. [NASA 1986] A subsequent, more detailed analysis yielded an estimate of 1 in 145 missions. [SAIC 1995] Of the 124 missions flown to date (as of mission STS-124), two have resulted in catastrophes.

U.S. Nuclear Safety Development.   The first use of event trees/fault trees for a major QRA in the nuclear industry was the Reactor Safety Study in 1975. [NRC 1975]   Since the 1980s, the Nuclear Regulatory Commission (NRC) has increasingly guided the commercial nuclear industry toward using QRA techniques to assess the industry's risk to the public. In 1986, NRC staff began using quantitative measures of risk to accomplish the Commission's safety goals. [NRC 1986]

In the early 1990s, the NRC began requiring each nuclear plant licensee to conduct an Individual Plant Examination of External Events (IPEEE), which quantified the risk associated with all analyzed incidents. Beginning in 2000, the NRC began implementing the philosophy of risk-informed decision making based on IPEEE analyses. Risk criteria have been established for proposed changes in the plant design. [Pfitzer 2007]   While nuclear reactor incidents can have immediate health effects, nuclear QRAs and their associated risk criteria are often complicated by the need to address the longer term, latent effects of radiation exposures.

Other Industries.   Examples of the use of risk criteria in civil engineering and transportation engineering applications also exist.   Other examples include (1) the establishment of earthquake design standards, which reference the likelihood of various levels of seismic activity, and set risk-based criteria to address the construction of structures to survive these earthquakes and (2) recent initiatives by the International Maritime Organization (IMO) for the use of risk-based approaches in evaluating the adequacy of ship design. Such applications are further exemplified in the discussion of risk criteria precedents in Appendix B.

2.1.3   Risk Assessment in the Process Industries

Early QRAs for Chemical Facilities.   The UK Canvey Island study and the Dutch Rijnmond study are early examples of QRA in the process industries. The former study, conducted in the 1970s to assess the risk of oil, liquefied gas, and chemical facilities located at the Canvey Island complex, was the most comprehensive risk assessment to date of non-nuclear installations in the United Kingdom.   The report was accepted by the UK Health and Safety Executive (HSE) as a "significant step forward in QRA, with some acknowledged deficiencies." [HSE 1978]   In 1981, a second Canvey Island report was issued, addressing the cited deficiencies and updating the risk estimates based on risk-reduction improvements at the facilities and improvements in the ability to model the risk. [HSE 1981]

The Rijnmond study was initiated by a Dutch commission in the early 1980s to evaluate the risk of various petrochemical storage and processing sites in a heavily industrialized area of the Netherlands[2]. The study was one of the first major QRAs performed by the chemical industry and is still referenced in current studies. The study selected six clearly defined installations and evaluated the frequency and consequences of possible releases of toxic and/or flammable material from each installation. Fairly detailed frequency and consequence modeling techniques were used to develop the risk estimates for the various sites. [Rijnmond 1982]

Other early uses of risk assessment included:

- Evaluation of proposed liquefied natural gas (LNG) import projects in California in the 1970s. Western LNG Terminal Company commissioned risk assessments, which projected low safety risks. The projects were eventually cancelled, not due to risk concerns, but because of an increasing domestic natural gas supply that made the projects uneconomical at the time. [CEC 2003]
- A QRA of the shipment of ten substances (acrylonitrile, anhydrous ammonia, benzene, caustic soda, chlorine, ethylene glycol, methanol, styrene, sugar, and sulphuric acid) between specific locations within the US. Each commodity was analyzed for safety risk for shipment by barge, truck, rail, and pipeline (as appropriate) in terms of quantitative estimates of the expected annual number of fatalities for each substance for each mode of transport. [ADL 1974]

Current Use of Risk Criteria. Either implicitly or explicitly, the process industries are increasingly seeking to make risk-informed decisions regarding their operations. Many major corporations are using quantitative risk assessment tools as part of an integrated program to evaluate and manage their risks. As suggested in Chapter 1, risk criteria are useful in interpreting the significance of the results of such quantitative studies and in making informed judgments on reducing risks.

At the same time, regulatory authorities in some countries are increasingly mandating the conduct of QRA for objectives such as: (1) planning approval for new chemical or petroleum facilities, (2) planning approval for new off-site developments in the vicinity of existing chemical or petroleum facilities, (3) safety cases in support of the renewal of operating licenses, and (4) assessment of proposed hazardous material transportation routes. Relevant precedents will be summarized in Chapter 3 and Appendix B.

---

[2] The Commissie Veiligheid Omwonenden (COVO) (commission for safety of inhabitants) of the Rijnmond Authority.

2.2   THE QUALITATIVE APPROACH TO RISK ASSESSMENT

Chapter 1 introduced the concept of risk as a function of both the undesired consequences and the likelihood of an unexpected event. It also noted that the consequence and frequency estimates may be based upon subjective, qualitative judgments or may be determined through quantitative modeling. While focused on the use of risk criteria in QRA, these *Guidelines* will further discuss qualitative approaches to broaden the understanding of the application of risk criteria in QRA.

2.2.1   PHAs as an Example of Qualitative Risk Assessments

One of the more common applications of qualitative risk assessment is in the performance of process hazard analyses (PHAs). These studies seek to identify and evaluate the significance of potential incident scenarios using such techniques as: what-if analysis, checklist analysis, failure modes and effects analysis (FMEA), or hazard and operability analysis (HAZOP). [CCPS 1992] Regardless of the technique selected, the analyst or review team seeks to answer the six fundamental questions detailed in Section 1.1.2 to gain perspective on the nature, magnitude, and tolerability of risks associated with a process.

> **Addressing Other Risks**
>
> Risk matrices can also be generalized to address many diverse types of risk, such as:
> * Equipment damage
> * Business interruption costs
> * Environmental damage
> * Loss of reputation

Commonly, these questions are addressed subjectively in a PHA. Gradations of consequence and likelihood (e.g., Low, Medium, High) are addressed either implicitly (based upon the individual perceptions of each team member) or explicitly (based upon definitions shared with the team). Judgments about consequences, likelihood, and the tolerability of the resulting risk are often made on a subjective basis using the collective knowledge and experience of the PHA team members and may not be consistent from team to team across the organization, or even within the same study conducted by the same team.

The Risk Matrix – Enhancing Qualitative Risk Assessment. PHA teams are increasingly using a risk matrix, such as the one shown in Figure 2.1, to help promote more consistent qualitative risk decision making and to help document the results. The two scales on the matrix describe increasing levels of consequence and frequency. The cells of the matrix define, for each consequence/frequency pair, a relative degree of risk. For this example, four levels of risk are identified (an increasing level of risk from IV through I).

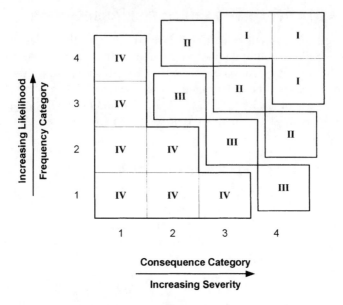

FIGURE 2.1. Example Qualitative Risk Matrix [CCPS 1992]

The example qualitative risk matrix shown in Figure 2.1 uses consequence categories for either the public or employees that are defined in Table 2.1 in a qualitative manner (such as no injury, minor injury, severe injury, fatality, etc.). The frequency categories described in Table 2.2 are also in a qualitative form (e.g., not expected to occur during life of process/facility).

TABLE 2.1. Example Consequence Categories – Qualitative Risk Matrix

Category	Description
1	No injury or health effects
2	Minor to moderate injury or health effects
3	Moderate to severe injury or health effects
4	Permanently disabling injury or fatality

TABLE 2.2. Example Frequency Categories – Qualitative Risk Matrix

Category	Description
1	Not expected to occur during life of process/facility
2	May occur once during life of process/facility
3	May occur several times during life of process/facility
4	Expected to occur more than once in a year

An example of risk ranking categories typical for a qualitative risk matrix is displayed in Table 2.3. Required responses are given for the four levels of risk displayed in the risk matrix illustrated in Figure 2.1 (from *Unacceptable* for the highest risk level to *Acceptable As Is* for the lowest). Often, the underlying risk criteria (if such exist) may not be discernible from such a qualitative risk matrix.

TABLE 2.3. Example Risk Ranking/Response Categories – Qualitative Risk Matrix

Risk Level	Description	Required Response
I	Unacceptable	Immediate mitigation or termination of activity
II	High	Mitigation within 6 months
III	Moderate	Mitigation within 12 months
IV	Acceptable As Is	No mitigation required

2.2.2   Semi-Quantitative Risk Matrices

The qualitative risk matrix presented in Figure 2.1 is typical for applications in which judgments about frequencies and consequences are purely subjective (i.e., based upon the team's 'gut feel'). While quantitative analysis may provide additional insight that may not be apparent in a purely qualitative analysis, a QRA may not be warranted for some studies.

The typical next step toward quantification of risk is an approach such as that exemplified by the risk matrix provided in Table 2.4. This particular example is taken from the military standard for system safety. [DOD 2000] Tables 2.5 and 2.6 are the consequence and frequency

> **Semantics**
>
> Some people would assert that, any time numbers are involved, the risk assessment is *quantitative* – the only issue is the rigor going into creating the numbers. While this logic might be sound, the term *semi-quantitative risk matrix* is used to address this topic here to consolidate the discussion of matrix-based approaches.

category definitions for this matrix. This risk matrix approach is termed semi-quantitative in that numerical definitions are provided for the consequence and frequency gradations. However, it remains only semi-quantitative because consequence and/or frequency modeling is not necessarily used in estimating consequence and frequency. Nevertheless, many find that this approach is of greater value to an experienced PHA team since, based upon its collective experience, the team has at least a historical sense of how frequently an event might occur or how great the consequences might be.

TABLE 2.4. Example Semi-Quantitative Risk Matrix [DOD 2000]

Severity Probability	Catastrophic	Critical	Marginal	Negligible
Frequent	1	3	7	13
Probable	2	5	9	16
Occasional	4	6	11	18
Remote	8	10	14	19
Improbable	12	15	17	20

TABLE 2.5. Suggested Mishap Severity Categories for Semi-Quantitative Risk Matrix [DOD 2000]

Description	Category	Environmental, Safety, and Health Results Criteria
Catastrophic	I	Could result in death, permanent total disability, loss exceeding $1M, or irreversible severe environmental damage that violates law or regulation.
Critical	II	Could result in permanent partial disability, injuries or occupational illness that may result in hospitalization of at least three personnel, loss exceeding $200K but less than $1M, or reversible environmental damage causing a violation of law or regulation.
Marginal	III	Could result in injury or occupational illness resulting in one or more lost work day(s), loss exceeding $10K but less than $200K, or mitigatible environmental damage without violation of law or regulation where restoration activities can be accomplished.
Negligible	IV	Could result in injury or illness not resulting in a lost work day, loss exceeding $2K but less than $10K, or minimal environmental damage not violating law or regulation.

One can assert that a disparity often exists in the comprehension of frequency estimates and consequence estimates. The typical PHA member can likely appreciate the contrast in the consequences potentially associated with a small fire versus a catastrophic vapor cloud explosion – most team members will be calibrated by either their own personal experiences or their awareness of other events that have occurred elsewhere within industry. However, gaining an intuitive appreciation of frequency is much more difficult when the team, reviewing a process that may be only 20 years old, seeks to discriminate between events that are predicted to occur once every 100 years versus once every 1000 years versus once every 10,000 years.

TABLE 2.6. Suggested Mishap Probability Levels for Semi-Quantitative Risk Matrix [DOD 2000]

Description*	Level	Specific Individual Item	Fleet or Inventory**
Frequent	A	Likely to occur often in the life of an item, with a probability of occurrence greater than 10^{-1} in that life.	Continuously experienced.
Probable	B	Will occur several times in the life of an item, with a probability of occurrence less than 10^{-1}, but greater than 10^{-2} in that life.	Will occur frequently.
Occasional	C	Likely to occur sometime in the life of an item, with a probability of occurrence less than 10^{-2} but greater than 10^{-3} in that life.	Will occur several times.
Remote	D	Unlikely but possible to occur in the life of an item, with a probability of occurrence less than 10^{-3} but greater than 10^{-6} in that life.	Unlikely, but can reasonably be expected to occur.
Improbable	E	So unlikely it can be assumed that occurrence may not be experienced, with a probability of occurrence of less than 10^{-6} in that life.	Unlikely to occur, but possible.

* Definitions of descriptive words may have to be modified based on the quantity of items involved.
** The expected size of the fleet or inventory should be defined prior to accomplishing an assessment of the system.

The risk levels represented in the risk matrix displayed in Table 2.4 can be interpreted as requiring an increasing level of required response. Note that this specific matrix has a more refined scale of risk categories (1 through 20). These categories are used to categorize and specify tolerance levels. The organizational titles shown in Table 2.7 indicate that continued acceptance of increasingly higher levels of risk must be sanctioned at increasingly higher levels in the organization. A similar approach is often used in the process industries, where it is sometimes termed *risk escalation*.

TABLE 2.7. Example Mishap Risk Categories and Mishap Risk Acceptance[‡] Levels [DOD 2000]

Mishap Risk Assessment Value	Mishap Risk Category	Mishap Risk Acceptance Level[‡]
1 – 5	High	Component Acquisition Executive
6 – 9	Serious	Program Executive Officer
10 – 17	Medium	Program Manager
18 – 20	Low	As directed

[‡] The term 'Acceptance Level' (rather than tolerance level) was used to be consistent with the original source of this table.

The increasing level of authorization associated with higher levels of risk implies some underlying risk criteria. While inferring the risk criteria from the information presented might be difficult, risk decisions driven by such a matrix should be consistent with other applications of the risk criteria established by the organization. This goal of consistency is complicated by the fact that risk matrices are applied to individual scenarios while company risk criteria typically address the integrated risk of operations. This issue is discussed further in Section 4.12.

2.2.3   Other Uses for Risk Matrices

PHAs are not the only activity in which a matrix-based approach to risk decisions has been used. Other examples include:

- Risk-based prioritization of maintenance and inspection activities
- Risk-based prioritization of follow-up activities for incident investigation or audit recommendations
- Determination of the required level of rigor in an incident investigation, based upon the perceived risk of a recurrence of the subject incident
- Risk-based determination of the required rigor of review or level of authorization for management of change requests
- Specification of the required risk reduction for an instrumented protective system (IPS)

2.2.4   A Brief Comparison of Qualitative and Quantitative Risk Assessment

Table 2.8 contrasts the application of qualitative and quantitative approaches to risk assessment. This is a partial list of attributes and is not intended to be comprehensive. Existing Center for Chemical Process Safety (CCPS) documents, such as *Guidelines for Chemical Process Quantitative Risk Analysis (CPQRA Guidelines)* [CCPS 2000a] and *Evaluating Process Safety in the Chemical Industry – A User's Guide to Quantitative Risk Analysis (QRA*

User's Guide) [CCPS 2000b], imply that, as the predicted consequences of incidents increase, the level of analytical detail should also increase (i.e., from strictly qualitative analyses toward detailed QRA).

The spectrum of risk assessment shown in Figure 2.2 illustrates the concept of performing multiple layers of assessment. Each layer provides more detailed information and reduces the uncertainty associated with results, but requires greater resources. Analysis is typically performed at the level necessary to obtain the appropriate information for risk-based decisions. This concept is further illustrated in Figure 2.3. For all chemical and petroleum facilities, some type of hazard identification should be performed. At a certain threshold of consequence, hazard evaluation studies are typically performed. As the predicted consequences become more severe, more detailed QRA may be appropriate.

2.3   TECHNICAL ASPECTS OF QRA

As discussed in the *CPQRA Guidelines* [CCPS 2000a], QRA is a methodology designed to give management a tool for evaluating overall process risk. Other aspects of risk management, such as the implementation of a risk-based process safety management system as described in CCPS' *Guidelines for Risk Based Process Safety* (*RBPS Guidelines*) [CCPS 2007a], may provide layers of protection against process incidents. However, the potential for serious incidents cannot be eliminated. QRA provides a quantitative method to evaluate risk and to identify areas for effective risk reduction.

A basic understanding of QRA methodology may be of value in helping to understand the application of risk criteria. The following text has been drawn from the *CPQRA Guidelines* [CCPS 2000a], with some updating and elaboration. Figure 2.4 illustrates the basic steps in a QRA.

2.3.1   Consequence Modeling

The incidents of concern within the process industries are often, but not always, associated with the release of materials with hazardous properties, such toxicity and energy content (e.g., thermal, pressure, or potential combustion energy). Typical incident scenarios include the rupture or break of a pipeline, a hole in a tank, a runaway reaction in a vessel, fire external to the vessel causing a relief valve to open, an operator erroneously opening a vent or drain valve, and so forth.

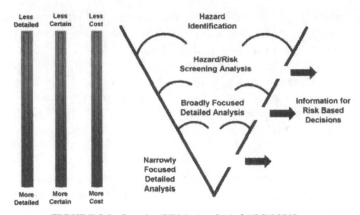

FIGURE 2.2. Levels of Risk Analysis [ABS 2000]

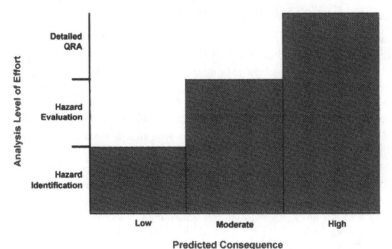

FIGURE 2.3. Scale of Risk Assessment Efforts

Once the scenario is defined, a source model(s) is selected to describe how materials are discharged from the process. The source model describes the rate of discharge, the total quantity discharged (or total time of discharge), and the state of the discharge – that is, solid, liquid, vapor, or a combination. Typically, a dispersion model is subsequently used to describe how the material moves downwind and mixes with air to some concentration.

For toxic releases, effect models consider the concentration and the duration of exposure and convert these incident-specific results into estimates of the effects on people (injury or death). For flammable releases, fire and explosion models consider information on the concentration and mass of

TABLE 2.8. Attributes of Qualitative Versus Quantitative[‡] Studies

Attribute	Qualitative Analysis	Quantitative Analysis
Scope	• Useful in assessing risk in a wide variety of applications • Easily customized to new applications • Difficult to apply for high-consequence, low-frequency risks	• Can provide estimate of "total" risk of an enterprise • Provides additional information, based upon science and technology, for decision makers to use in identifying and implementing appropriate risk mitigation measures • Helps ensure that facilities with very low probability for serious incident events are designed to meet the company's risk criteria. • Useful for addressing high-consequence, low-frequency events • Typically reserved for "as-needed" situations • General guidance is to use just enough quantification to answer the risk-based questions
Ease of use	**Benefits –** • Quick and easy to use • Easy to train new personnel • Easier to explain to the consumer of the study **Limitations –** • Potential for inconsistent use of risk matrices • Risk decisions are often made based upon the experience of practitioners with much less risk assessment background • Tendency to answer questions based upon the user's personal experience, rather than considerations of what could happen	**Benefits -** • Prescribed techniques exist for performing QRA **Limitations –** • Requires specialized expertise • More time consuming and requires greater resources to conduct • Depends upon the availability of reliable models and failure rate data

23

TABLE 2.8. Attributes of Qualitative Versus Quantitative[‡] Studies

Attribute	Qualitative Analysis	Quantitative Analysis
Objectivity	**Benefits –** • Methodologies can be tailored to meet the needs of the study **Limitations –** • Risk evaluations are subjective • Generally applied on a scenario-by-scenario basis and difficult to compare with QRA results • Difficult to place an event with many possible outcomes within a single box (e.g., for a toxic release, consequences depend on wind direction and weather conditions at the time). • Difficult to standardize frequency definitions (i.e., is the frequency associated with a single piece of equipment, process unit, or site) • Difficult to address the reliability of safeguards such as safety instrumented systems or mechanical integrity programs	**Benefits –** • Results can be more reproducible, if methodologies and databases are standardized (see Section 2.3.4) • The level of detail required adds to the understanding of the process and provides a better historical record of the decisions • The effectiveness of risk-reduction alternatives can be compared **Limitations –** • Detail involved can make QRAs harder to understand • QRA should only be used where reliable data supports the study • Only qualified analysts can make appropriate judgments as to the effects of specific uncertainties on decision making

‡ As described in the sidebar in Section 2.2.2, semi-quantitative methods bridge between the qualitative and quantitative methods.

CPQRA Steps

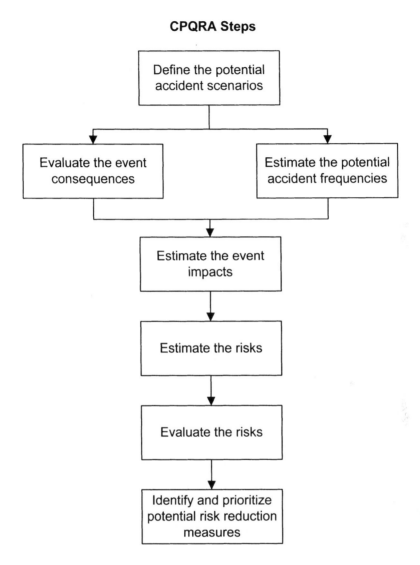

FIGURE 2.4. QRA Process [CCPS 2000a]

material present in the flammable cloud (and, perhaps, information describing the physical environment, such as the confinement and congestion of the flammable cloud) to estimate energy hazard potentials, such as thermal radiation and explosion overpressures. Other effect models are then used to estimate effects on people and structures.

Additional refinement to consequence estimates may be provided by considering mitigation factors, such as isolation systems that might reduce the duration of the release, dikes that limit the spread of liquid spills, use of foam to limit evaporation, and sheltering or evacuation that may reduce the magnitude of potential effects.

Figure 2.5 shows a flowchart illustrating some of the considerations that go into modeling the consequences of the types of events discussed above. Neither this section, nor the figure, is intended to provide a comprehensive discussion of the types of events that may be encountered or the models that may be used. For example, bursting pressure vessels would not require dispersion modeling to estimate the resulting explosion overpressure. The reader should refer to the *CPQRA Guidelines* [CCPS 2000a], CCPS' *Guidelines for Evaluating the Characteristics of Vapor Cloud Explosions, Flash Fires, and BLEVEs* [CCPS 1994], and CCPS' *Guidelines for Evaluating Process Plant Buildings for External Explosions and Fires* [CCPS 1996] for additional guidance on consequence modeling.

2.3.2   Frequency Modeling

The *CPQRA Guidelines* [CCPS 2000a] provide detailed information on the most common techniques to answer the question, "How likely is this incident scenario to occur?" The main frequency analysis techniques are:

- Review of historical records of similar events
- Fault tree analysis
- Event tree analysis
- Layer of protection analysis
- External event analysis
- Common cause failure analysis
- Human reliability analysis

In performing frequency analyses, determining the appropriate level of detail needed to provide sufficient information to make the necessary risk-based decision is often difficult. [Farquharson 2003]   Typically, a phased approach, such as the following, should be considered in performing frequency analysis:

1.   Perform a qualitative study using, for example, HAZOP or What-if Analysis to identify potential initiating events that could lead to incident scenarios of interest.

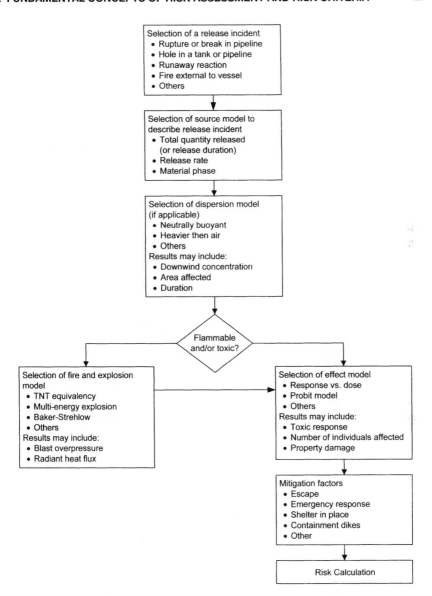

FIGURE 2.5. Consequence Analysis Flowchart [CCPS 2000a]

2. For initiating events of interest, prepare an event tree to further develop the scenarios; for example, showing the various outcomes that could result based upon the success or failure of relevant protective features.
3. Use techniques such as fault tree analysis or the review of historical records to estimate the initiating event frequencies and branch point probabilities for each scenario.
4. Calculate the frequency estimates for each scenario outcome by multiplying the initiating event frequency by the appropriate branch point probabilities.

Figure 2.6 illustrates a hypothetical example of an event tree showing estimates made for the initiating event frequency and the branch point probability levels. Simplified frequency analyses can be performed by estimating branch point probabilities without conducting a detailed fault tree analysis.

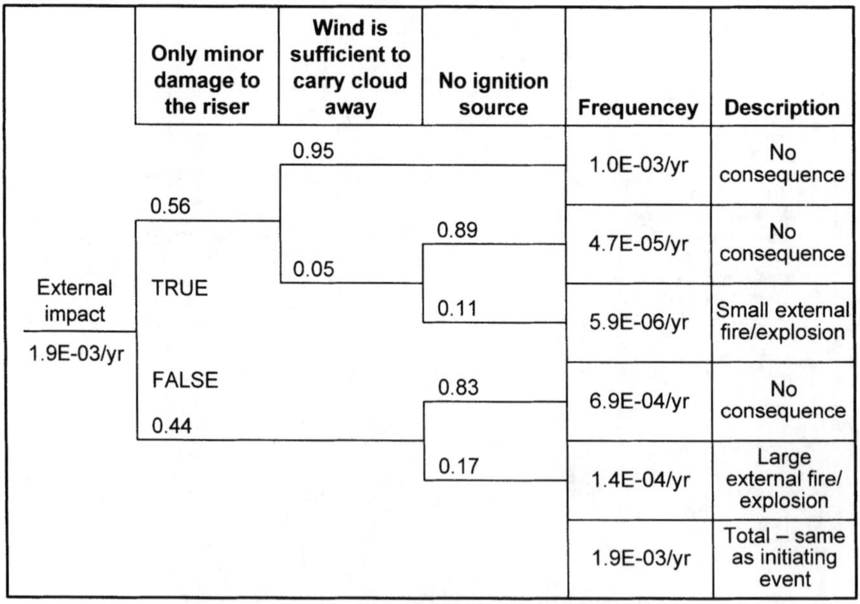

Only minor damage to the riser	Wind is sufficient to carry cloud away	No ignition source	Frequencey	Description
0.95			1.0E-03/yr	No consequence
0.56				
	0.89		4.7E-05/yr	No consequence
	0.05			
TRUE		0.11	5.9E-06/yr	Small external fire/explosion
	0.83		6.9E-04/yr	No consequence
FALSE				
0.44				
	0.17		1.4E-04/yr	Large external fire/ explosion
			1.9E-03/yr	Total – same as initiating event

External impact 1.9E-03/yr

Note: This figure is a hypothetical example and does not recommend values to be used for risk assessment purposes.

FIGURE 2.6. Sample Event Tree [Farquharson 2003]

Layer of protection analysis (LOPA) [CCPS 2001] is a more recent technique developed for analyzing the frequency of a single cause/consequence pair. As such, LOPA is a simplified form of risk assessment, using order of magnitude categories for initiating event frequency, consequence severity, and the likelihood of failure of independent protection layers (IPLs) to approximate the risk of a scenario. LOPA typically builds on the information developed during a qualitative hazard evaluation, such as a PHA. The primary purpose of LOPA is to determine if sufficient IPLs are in place to guard against an incident scenario (i.e., can the risk be tolerated?).

If the estimated risk of a scenario is not tolerable, additional IPLs may be specified. LOPA does not suggest which IPLs to add or which design to choose, but it assists in judging between alternatives for risk mitigation. LOPA is not a fully quantitative risk assessment approach, but is rather a simplified method for assessing the value of protection layers for well-defined incident scenarios.

2.3.3 Developing a Comprehensive QRA

Terms like *worst case scenario* or *worst credible scenario* often creep into discussions related to risk assessments. Every scenario has a frequency and consequence (and, therefore, risk) associated with it, and the significance of that scenario cannot be determined until the risk has been quantified (or at least estimated). As mentioned earlier, most high-consequence incident scenarios occur at a relatively low frequency. Examples of events that are normally categorized as "worst credible scenarios" include full-bore ruptures of liquid lines and pressure vessel failures. Whether or not these incident scenarios may actually be the worst from a consequence standpoint, their risk significance would not be clear unless a full spectrum of typical incident scenarios is assessed (i.e., including more moderate consequence events that are more likely). For example, a medium size pipe rupture that results in a vapor cloud explosion may have a lower consequence compared to a pressure vessel rupture, but the risk may be determined to be higher once all scenarios are quantified. For this reason, these *Guidelines* recommend that a representative range of identified scenarios should be evaluated in a QRA. Scenarios should not be excluded based upon an *a priori* decision of, "That's too unlikely to consider."

Readers may wish to refer to page 22 and Figure 1.8 in the *CPQRA Guidelines* [CCPS 2000a] for a more detailed discussion of the identification of scenarios for consideration in a QRA.

The above notwithstanding, some types of incidents may exist whose likelihoods are beyond current ability to predict. For this reason, and as an example, analysts often identify the consideration of terrorist threats as being beyond the scope of a typical QRA. The report documenting a QRA study should identify any such events that have been declared to be beyond the scope of the study.

2.3.4   Standardization of Approach

Consequence and frequency modeling are complex and evolving topics and, often, divergent opinions can exist, even among the experts, as to the best way to model a particular scenario. In fact, past comparative studies involving multiple independent analysts modeling the same scenario have yielded results that range from one to two orders of magnitude from the mean. However, once the teams were coached to use similar assumptions, the risk results converged to within an acceptable range (i.e., within a factor of 5). [CCPS 2000a]

Certainly, if an organization commits to the use of QRA, well-defined guidelines regarding assumptions and techniques need to be established to promote consistency. Companies are increasingly striving to standardize across all locations so that any company-wide risk-based decision making is less likely to become an exercise in comparing apples to oranges. Corporate QRA guidance could provide standardized data and analytical approaches addressing topics such as:

- Equipment failure rate data
- Human error rates
- Toxicity dose-response relationships
- Physiological response to thermal exposures (fires) and explosion overpressure
- Structural analyses to determine building damage in response to explosions, and resultant occupant vulnerability
- Analytical techniques for source term and dispersion modeling
- Assumptions made regarding credit to be given for mitigative design features/activities, such as sheltering-in-place, remote isolation of leaks, water spray systems, and so forth.
- Selection of modeling software
- Training/qualification requirements for QRA analysts

Similarly, in some jurisdictions where QRA is mandated, regulatory authorities may require and implement such standardization through the prescription of standardized protocols (perhaps including scenario definitions) to minimize variability of results for similar situations between different organizations and analysts. For example, the Netherlands has developed a suite of modeling software, including embedded data on failure rates. [DNV 2007] As of October 2008, companies that fall under the Dutch BRZO regulation (see section 3.3.6) are required to use this software (i.e., Safeti-NL).

2.4   QUANTITATIVE RISK CRITERIA

Before beginning a detailed discussion of risk criteria, it is necessary to define the risk measures to which the risk criteria apply. Experience has shown that, to get a balanced perspective of the risks associated with process plant operations, risks must be evaluated from two perspectives: (1) the risk to individuals and (2) the risk to groups of people. These are referred to, respectively, as individual and societal risk.

Many diverse measures of individual and societal risk exist. Those addressed here are the more commonly applied in the process industries. Readers seeking a broader perspective may wish to consult the *CPQRA Guidelines* [CCPS 2000a] for other examples of risk measures and formats for their presentation.

2.4.1   Individual Risk

Individual Risk expresses the risk to a single person exposed to a hazard; in other words, an individual in the potential effect zone of an incident or set of incidents. The scale of any incident, in terms of the number of people impacted by a single event, does not affect individual risk. Individual risk measures can be single numbers, tables of numbers, or various graphical summaries. Commonly used individual risk measures include [CCPS 2000a]:

1. *Individual risk contours* show the geographical distribution of individual risk. The risk contours are calculated from the expected frequency of an event capable of causing the specified level of harm at a specified location, regardless of whether or not anyone is present at that location to suffer that harm. Thus, individual risk contour maps are generated by calculating individual risk at every geographic location assuming that somebody will be present, unprotected (e.g., outdoors), and subject to the risk 100% of the time (i.e., annual exposure of 8,760 hours per year). In contrast, the other risk measures described below consider the fraction of the time that the individual is exposed to the risk. See Figure 2.7 for an example of an individual risk contour. An *individual risk profile,* as illustrated in Figure 2.8, is a plot of individual risk as a function of distance from the risk source and is a simplification of a risk contour. The risk depicted in these figures is associated with a particular location rather than a particular person. For this reason, this risk measure is sometimes referred to as *location risk* or *geographical risk.*

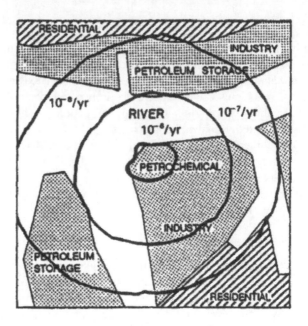

FIGURE 2.7. Example Individual Risk Contour [CCPS 2000a]

2. *Maximum individual risk* is the individual risk to the person exposed to the highest risk in an exposed population. For example, this might be the operator working at the unit being analyzed or, for a normally unstaffed location, it might be the person in the general population living at the location of highest risk. Maximum individual risk can be determined by calculating individual risk at every geographical location where people are present and searching the results for the maximum value. Alternatively, this calculation may seek the greatest risk considering individual working patterns (i.e., time spent on different activities, in different locations).

3. *Average individual risk (exposed population)* is the individual risk averaged over the population that is exposed to risk from the specified scenarios (e.g., all of the operators in a building, or those people within the largest incident effect zone). This risk measure is useful only if the risk is relatively uniformly distributed over the population, and can be extremely misleading if risk is not evenly distributed. For example, a few individuals may be exposed to a very high risk, but this fact may not be apparent when their risk is averaged with a large number of people at lower risk.

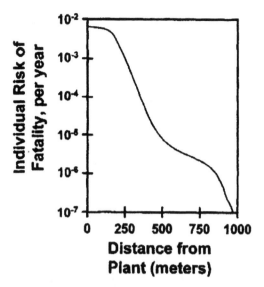

FIGURE 2.8. Example Individual Risk Profile [CCPS 2000a]

4. *Average individual risk (total population)* is the individual risk averaged over a predetermined population, without regard to whether or not all people in that population are actually exposed to the risk. This average risk measure is potentially extremely misleading. If the population selected encompasses many individuals who are exposed to little or no risk, the average individual risk will obscure the risk to those at greatest risk.

5. *Average individual risk (exposed hours/worked hours)* is the individual risk for an activity that is calculated for the duration of the activity or that is averaged over the working day. For example, if an operator spends 1 hour per shift sampling a reactor and 7 hours per shift in the control room, the individual risk while sampling would be 8 times the average individual risk for the entire work day, assuming no risk for the time in the control room.

Other bases for calculating individual risk have been used. Considering the multiplicity of individual risk measures, ensuring consistency between the manner in which individual risk is calculated and the basis upon which the risk criteria are defined is important (See Section 3.5.6).

The individual risk measures described above are normally expressed as the frequency of fatal injury per year. While all injuries are of concern, effect models for predicting degrees of injury often include additional uncertainties; thus, risk analysts often estimate the risk of fatal injury (death) as a less equivocal measure. As will be noted later, more comparative historical data are also available that can be used to calibrate the risk criteria against the risk of death associated with other hazardous activities.

Other individual risk measures that have been used include the *Fatal Accident Rate* (FAR). [Lees 2004] This is the estimated number of fatalities per 10^8 exposure hours (roughly equal to the cumulative number of working hours over the lifetimes of 1,000 employees). The FAR is a single number index that is directly proportional to the average individual risk. To calculate the FAR, multiply the average individual risk by a factor of $10^8/(24 \times 365) = 1.14 \times 10^4$.

Individual risk can be calculated for either on-site personnel (e.g., employees or contractors) or for members of the off-site public. Calculations of public risks can introduce considerations that are either unique or more severe than for on-site risk calculations. For example:

- Off-site populations may include individuals who, because of their age or general health condition (e.g., young children or the elderly), may be more vulnerable to health impacts and may be less capable of responding to protect themselves.
- Off-site individuals are less likely to be aware of process hazards and the means to protect themselves from those hazards and, most likely, will not have the same sorts of protective equipment that on-site personnel would have.
- Off-site individuals, while perhaps further removed from the hazards, may be exposed for a greater percentage of the time than are workers (e.g., stay-at-home residents who may be at risk nearly 100% of the time). Conversely, depending upon the nature of off-site developments, some individuals' risk exposure may be transient and brief (e.g., visitors to a park adjacent to a chemical or petroleum facility).

The calculation of *individual risk* is made with the understanding that the contributions of all incident outcome cases (i.e., event sequences) are additive. For example, the total *individual risk* to an individual working at a facility is the sum of the risks from all potentially harmful incidents considered separately, in other words, the sum of all risks due to fires, explosions, toxic chemical exposures, and so forth, to which the individual might be exposed.

2.4.2   Societal Risk

Some major incidents have the potential to affect large numbers of people; *societal risk* expresses the cumulative risk to groups of people who might be affected by such events.   In other words, *societal risk* measures the potential for impacts to a group of people located in the effect zone of an incident or set of incidents. Thus, societal risk estimates include a measure of incident scale in terms of the number of people impacted.   Some societal risk measures are designed to reflect the observation that society tends to be more concerned about the risk of large (multi-fatality) incidents than

> **Aggregate Risk**
>
> The American Petroleum Institute (API) and CCPS have both used the term *aggregate risk* to address the special case of *societal risk* for on-site personnel (particularly personnel inside of buildings). [API 2003 and CCPS 1996] Since this is just a special case of *societal risk*, and is calculated in the same fashion, in these *Guidelines* the term *societal risk* is used to address both on-site and off-site populations.

small (fewer fatality) incidents, and may assign greater significance to large incidents.   This potential risk aversion will be discussed further when addressing risk criteria formulation.

Societal risk measures can be expressed as single number measures, tabular sets of numbers, or graphical summaries, with the most common graphical representation being the Frequency-Number (F-N) curve.   An F-N curve is a plot of the frequency distribution of multiple-casualty events, where F is the cumulative frequency of all events leading to N or more casualties (typically expressed as the number of fatalities). F-N curves typically use log-log plots since the frequency and number of fatalities often range over several orders of magnitude.   Figure 2.9 shows examples of three F-N curves, and Appendix A illustrates the calculations required to develop an F-N curve.

The calculation of *societal risk* requires the same frequency and consequence information as *individual risk*. Whereas *individual risk* requires details of an individual's occupancy within hazard zones, *societal risk* estimation requires a definition of the number of exposed population within hazard zones. This definition can include factors such as (1) the number and geographical distribution of the population, (2) the population type (e.g., residential, school, industrial), and (3) the likelihood of people being present (i.e., including the number of hours a day people are present).

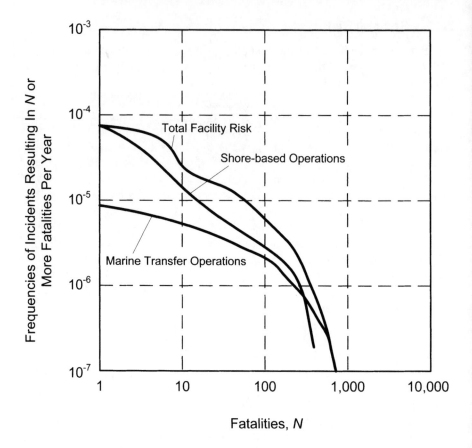

FIGURE 2.9. Example F-N Curve [CCPS 2000a]

Other societal risk measures that have been used include:

- *Average Rate of Death* [Lees 2004] is defined as the average number of fatalities that might be expected per unit time from all possible incidents. This measure is also known as the accident fatality number and is a single number average measure of societal risk.

- *HSE Risk Integral* [Hirst 2002] has been proposed by the UK HSE as a means of calculating a single value societal risk index. The risk integral can be calculated from the data used to construct an F-N curve. However, the advantage cited by the HSE is that the risk integral can be estimated based upon an assumed shape for the F-N curve and information on a proposed "worst case" scenario for the facility under study (i.e., the scenario having the highest number of fatalities).   These *Guidelines* outline concerns related to the identification of 'worst case scenarios" in Section 2.3.3.

Traditional emphasis has been on the calculation of societal risk for off-site populations; however, companies are increasingly recognizing the importance of the consideration of group risk for on-site personnel. As with individual risk, societal risk estimates are typically the summation of risk contributions from many incident outcome cases.

2.5   THE ROLE OF QRA AND RISK CRITERIA

"Begin with the end in mind" is a quote from author Steven Covey. [Covey 1990] Risk analysts can apply this advice in the realm of risk assessment by not calculating risk values until they know what will be done with the results. From the manager's perspective, the objective of a QRA is to provide information to decision makers to allow rational risk management decisions. Examples of the application of QRA include:

- Identifying major contributors to risk.   As noted above, both individual and societal risk estimates are commonly the summation of risk contributions from many scenarios.   The risks associated with a relatively few scenarios can dominate the sum.   Identifying and addressing the most significant contributors is an effective means of stewarding risk reduction resources.
- Comparing risk management alternatives. QRAs are often used to evaluate the risk reduction benefits of one alternative relative to others.   Options that represent the most prudent investment of risk reduction resources can be identified.
- Comparing risks of existing operations. A company may seek to compare the risk of several of its operations without making absolute judgments.
- Defining approval levels under the risk escalation principle. Organizations may require that higher risk activities be sanctioned at higher levels of authority within the organization.
- Making hard decisions about the tolerability of risk. QRAs can form the basis for absolute go/no-go decisions regarding a particular course of action or provide significant input into the deliberation of, "Have we done enough to reduce the risk?" (See Section 2.7, regarding the ALARP principle.)
- Achieving regulatory compliance. Some regulatory authorities may require QRAs to justify the initiation or continuation of certain hazardous activities.

Of the six general applications above, only the last three require the use of risk criteria (the first two promulgated by the company and the last promulgated by the regulatory authority). While this book is focused on the creation of risk criteria, the importance of those applications of QRA that require only relative evaluations of risk should be acknowledged. In fact, because of the uncertainties inherent in the assumptions made and the models used in QRA, some analysts may be more comfortable with comparative situations in which any inaccuracies apply, presumably equally, to all risk estimates so that relative judgments remain valid (see Section 2.8). However, in the absence of risk criteria, determining whether all (or none) of the alternatives being offered represent a tolerable risk is often difficult.

While not all risk management decisions require risk criteria, such criteria are essential for others – and such situations are becoming increasingly common. Some examples are cited below.

2.5.1 Impact of Evolving Consensus Standards and Recommended Practices

The process industries are experiencing increasing societal (and, in some places, regulatory) expectations for companies to more effectively manage the risks of their operations and facilities. Generally, however, the availability of risk reduction resources is not growing commensurate with these expectations. Standards-making bodies, trade associations, and technical societies are responding to these trends with new consensus standards and recommended practices that embody risk-based approaches to controlling hazards. Several examples of these standards and practices are described below.

Instrumented Protective Systems (IPS). The process industries rely upon many types of IPSs to respond to abnormal operations or situations that might otherwise have serious process safety consequences. Such systems are commonly identified as safeguards during PHAs and are classified based on the IPS type and the expected risk reduction. Ultimately, risk criteria are required to support the specification of the risk reduction requirements for each identified IPS. *Guidelines for Safe and Reliable Instrumented Protective Systems* [CCPS 2007b] discusses the lifecycle management of IPSs implemented to achieve a specified risk reduction strategy.

IPSs should be designed and managed to meet appropriate industry practices. For example, a safety instrumented system (SIS) is a type of IPS, which acts to achieve or maintain a process safe state when abnormal operation occurs. Consensus standards (such as ANSI/ISA 84.00.01 and IEC 61511) provide technical requirements for the design and management of an SIS to achieve a specified reliability based on four safety integrity levels.

Facility Siting Studies. Broadly, facility siting studies address the risk to building occupants posed by severe external challenges, such as the overpressure generated by an explosion or toxic threats. CCPS and the API have described risk-based approaches for evaluating the suitability of both building design and location to protect occupants from the risks associated with such challenges. [CCPS 1996 and API 2003]

Equipment Inspections. Mechanical integrity (MI) programs are intended, in part, to establish equipment inspection and testing frequencies to prevent failures that could lead to either loss of containment of process materials or loss of some safety-critical function. While PHAs can be helpful in identifying potentially safety-critical equipment, in the absence of risk criteria PHA teams tend to err on the side of conservatism and overstate the importance of equipment, potentially resulting in overburdened MI programs. Various standards, such as API RP-580, *Risk-Based Inspection,* [API 2002] provide risk-based approaches to identifying prudent inspection practices and frequencies.

Process Safety Management Systems. As mentioned in Chapter 1, the *RBPS Guidelines* [CCPS 2007a] describe a risk-based approach for identifying appropriate content and rigor of implementation for process safety management systems (after applicable regulatory requirements have been met). While many of the risk-based decisions required to implement the *RBPS Guidelines* may be made qualitatively, quantitative decisions requiring risk criteria may be appropriate in some situations.

2.6   RISK TOLERANCE AS A FUNCTION OF SOCIETAL VALUES

2.6.1   What Is 'Risk Tolerance'?

So far, these *Guidelines* have discussed risk tolerance without providing an explicit definition of *tolerance* or *tolerability*. In discussing the tolerability of risk, the UK HSE has written:

"... 'tolerable' does not mean 'acceptable.' It refers instead to a willingness by society as a whole to live with a risk so as to secure certain benefits in the confidence that the risk is one that is worth taking and that it is being properly controlled. However, it does not imply that … everyone would agree without reservation to take this risk or have it imposed on them." [HSE 2001]

As noted in Chapter 1, the concept of *risk tolerance* or *risk tolerability* is increasingly preferred to the concept of *risk acceptance*. The terminology has changed because organizations do not want to imply or create a public perception that risks attributable to their operations or activities are viewed as being acceptable. Rather, recognizing that eliminating all risks is impossible, some organizations prefer to speak of carefully managed residual risks being tolerable.

These *Guidelines* generally use the terms *risk tolerance* and *risk criteria*, except in cases in which the text references a particular organization that continues to use terms such as *risk acceptance* or *risk acceptance criteria* as their terminology of choice. Companies may elect to use different terminology.

2.6.2   Public Values as the Basis for Risk Criteria

Chapter 3 and Appendix B describe risk criteria precedents established by regulatory authorities and other organizations with, where pertinent, perspective on how these risk criteria were established. Key to several of these precedents is the concept that "the opinion of the public should underlie the evaluation of risk." [HSE 1992]   Whether explicitly or implicitly, the process industries receive 'permission to operate' from the public, and the public's level of confidence with respect to industry's management of risk must be a factor in the continued endorsement of industry activities.

As described in Chapter 3 and Appendix B, some regulatory authorities have published proposed risk criteria, seeking public scrutiny and commentary on the provisional criteria. A company may not be able to directly seek such input, especially larger companies operating in many communities or, even, countries. However, as an alternative, a company can infer the public's tolerance of risk from observations about the levels of risk that most people are willing to live with in respect to current patterns of employment, recreational activities, and public works that provide the benefits that society has come to rely upon. As will be noted, some of the most widely accepted risk criteria have been calibrated in this fashion.

Public values are subject to change, as evidenced by the tendency in the United States to become less tolerant of employment risks as the standard of living has increased and technology has advanced. As the ability to manage risks increases, anticipating that the public will expect industry to do better is reasonable. Consequently, risk criteria should be subjected to periodic reconsideration – probably not as frequently as yearly, but frequently enough to stay relevant and defensible.

Some common issues that help define the public's risk values are described in the following section.

2.6.3   Public Perceptions of Risk

Risk tolerability and risk perception are sensitive matters. While a variety of factors may influence a person's risk tolerance, the following are most relevant to defining risk criteria that "society as a whole can live with."

Scale of the Incident. For many people, risk tolerance is generally associated more with perceptions of the severity of the potential consequences of an incident than with the likelihood of its occurrence. For example, more public alarm is associated with an airliner crash killing 100 people than with the same number of fatalities occurring one or two at a time in automobile crashes. This is the case even though, for instance, several years might pass between such airline disasters in the U.S. – yet, each day, more than 100 highway fatalities occur in the U.S.

The construction of F-N curves for societal risk is described in Appendix A. Figure 2.10 plots an F-N risk criteria curve for the Netherlands for the general public (See Section 3.3). The curve shows that as the estimated number of fatalities from an event increases, the estimated frequency of the event must decrease accordingly to keep the risk within tolerable bounds (i.e., the risk must remain below the criterion curve shown in Figure 2.10). In other words, the more severe a potential incident is, the more unlikely its occurrence must be.

One defining feature of the risk criteria associated with an F-N curve is the slope of the curve. A curve with a slope of negative one (-1) indicates that, for every 10-fold increase in N, a corresponding 10-fold decrease in F must occur. The slope of the curve illustrates the degree of aversion to multi-fatality events. A slope of -1 is commonly described as being 'risk neutral' and, as the slope increases (e.g., from -1 to -2), an enhanced concern is expressed for events with the potential for multiple fatalities, expressing an aversion to large incidents.[3]

Tolerance of the Commonplace/Fear of the Unusual. People tend to become inured to those risks that they encounter on a routine basis, while remaining fearful of comparable risks for which they have no frame of

---

[3] Section A.2.2 describes that, while a slope more negative than -1 is often cited as being required to embody risk aversion in an F-N curve, the allegedly 'risk neutral' F-N curve (i.e., slope = -1) actually exhibits a significant degree of aversion to multi-fatality events.

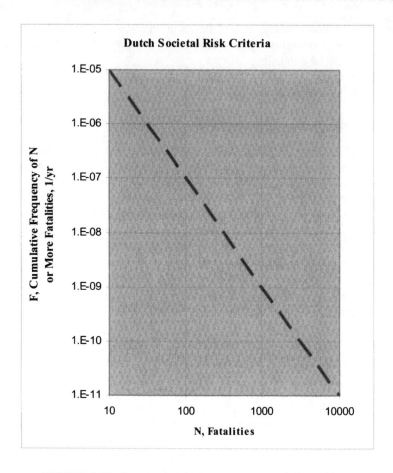

FIGURE 2.10. Current Dutch Societal Risk F-N Risk Criteria

reference.   The tolerance to "routine" highway deaths cited above is one example.   Another example might be the acceptance of an existing chemical or petroleum facility by the neighbors, while those same neighbors might express concerns regarding the construction of a new facility or the addition of a new process unit to an existing facility.

Control Over the Risk. Most people are generally willing to take on more risk if they perceive that they have some degree of

Further understanding of risk perception

Readers seeking a broader understanding of the factors that determine public perception of risk should refer to the work of Paul Slovic, such as [Slovic 1987]

control over that risk. For example, a person may be more comfortable at high speed in an automobile if he, rather than someone else, is driving. Or, a person may be more comfortable holding the nail if also holding the hammer.

Voluntary vs. Involuntary Risk. Closely related to the topic of control over the risk is the issue of whether the risk is taken voluntarily or whether the risk is imposed. Workers can, to a certain extent, choose their occupation. Consequently, workers are generally regarded to have voluntarily assumed the risk associated with their employment, in return for the benefits derived from that employment. This statement is particularly valid when the employer has taken reasonable and prudent measures to (1) control workplace hazards, (2) inform workers of the hazards, and (3) train workers in prevention and control of the hazards, in necessary personnel protection, and in emergency response actions.

Other risks voluntarily assumed in order to obtain some benefit include choosing air travel (to benefit from quicker transit to a destination) or engaging in a dangerous sport, such as mountain climbing or scuba diving (to benefit from the pleasure of the recreation).

In contrast, risks to members of the general public living in nearby residential areas or present in nearby schools, hospitals, or businesses around a chemical or petroleum facility are typically considered involuntary risks (i.e., risks imposed on the public). In other words, members of the general public have not necessarily assumed the industrial risks that are imposed upon them, even though individual members of the public may have voluntarily elected to purchase housing in the vicinity.

An important issue regarding the applicability of voluntary risk is the distinction of categories of personnel at an industrial site and how the risk criteria vary depending on the personnel. The different categories of personnel are sometimes defined as (1) process plant workers, (2) office workers, (3) on-site contractor workers, and (4) process workers from neighboring industrial sites. Making a distinction between process workers, office workers, and contract workers often becomes a difficult exercise with little benefit. Many companies are attempting to move non-essential personnel further away from the process units and developing exclusion zones in which non-essential personnel are not permitted during certain phases of operation, such as startup and shutdown, to reduce their risk. Workers located at neighboring facilities may be considered involuntary risk recipients (when considering the risk of the facility of interest), or voluntary risk recipients, or may be placed in a category between plant workers and involuntary risk recipients. Various precedents are discussed in Chapter 3 and Appendix B.

2.7   DEFINITION AND APPLICATION OF THE "AS LOW AS REASONABLY PRACTICABLE" (ALARP) PRINCIPLE

Figure 2.11 illustrates a *Tolerability of Risk* (TOR) model that is applicable to both individual and societal risk. [HSE 2001]   While originally developed by the UK HSE, this model has been widely adopted.   The model involves a three-tier action level scheme that classifies risk into the following regions:

- An *unacceptable* risk region, where further risk assessment or risk mitigation is required.   Activities whose risk cannot be reduced below this region are subject to termination.
- A *broadly acceptable* risk region where the risk is so low that further risk reduction need not be considered (but could be implemented on an opportunistic basis).
- Between these extremes is a region where the risk is deemed to be *tolerable* but where further, prudent risk reduction measures should be considered for implementation.

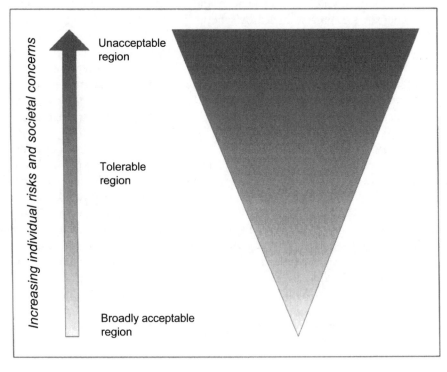

FIGURE 2.11. HSE Tolerability of Risk Diagram [HSE 2001]

Section 2.6.1 described the distinction between *risk acceptance* and *risk tolerance* and explained the rationale for avoiding the term *acceptable risk*. However, the HSE TOR model uses the term *broadly acceptable* to describe levels of risk that are "generally regarded as insignificant and adequately controlled" and that "are comparable to those that people regard as insignificant or trivial in their daily lives." [HSE 2001] These *Guidelines* use this term also, since it more emphatically distinguishes between risks that are *tolerable* (but should be considered for further reduction) and risks that are already so low that most would view further risk reduction efforts as being warrantless. HSE's use of the term *unacceptable* to describe excessively high risk introduces no potential for ambiguity.

This model suggests the creation of two risk criteria – an upper risk criterion defining the maximum tolerable risk, and a lower risk criterion defining broadly acceptable risk. Many companies and regulatory authorities have based their risk criteria on this model. Others use a modified 2-region/single risk criterion model that does not contain the lower, broadly acceptable risk criterion. This modified model distinguishes between risks that are unacceptable and risks requiring further, prudent reduction.

The 2-region model, in effect, implies a continuous improvement approach under which the company should remain alert to new opportunities to reduce risk, no matter how low the risk currently is. Implementation of such opportunities may not be obligatory, but they should be given fair consideration. As discussed in the next chapter, the Netherlands provides one example in which regulators have elected to replace a 3-region system with a 2-region system in order to stimulate such a search for continuous improvement.

Key to the implementation of either of these models is the matter of what constitutes prudent efforts. The UK has addressed this need with the concept of reducing risk *As Low As Reasonably Practicable* (ALARP). British courts have determined that the term *reasonably practicable* is a narrower term than *physically possible*. The UK HSE ALARP principle is interpreted and developed from a 1949 legal judgment establishing that, to assert ALARP:

> **Related Terms...**
>
> Terms such as *As Low As Reasonably Achievable (ALARA)* and *So Far As Is Reasonably Practicable (SFAIRP)* express concepts similar to *ALARP*. In some cases, regulatory distinctions between these terms exist. However, the UK and the Netherlands use, respectively, ALARP and ALARA to address the same concept.

"a computation must be made in which the quantum of risk is placed in one scale and the sacrifice involved in the measures necessary for averting the risk (whether in money, time or trouble) is placed in the other, and that, if it be shown that there is a gross disproportion between them - the risk being insignificant in relation to the sacrifice - the defendents discharge the onus upon them." [Lord Justice Asquith, quoted in RSSG, 1983]

These *Guidelines* propose the following definition for ALARP:

As Low As Reasonably Practicable (ALARP): The concept that efforts to reduce risk should be continued until the incremental sacrifice (in terms of time, effort, cost, or other expenditure of resources) is grossly disproportionate to the value of the incremental risk reduction achieved.

Extensive guidance on the ALARP concept has been issued by the UK. Some basic premises include: [HSE 2008a]

- By 'grossly disproportionate,' the HSE clearly intends that the 'sacrifice' must be significantly greater than the risk reduction achieved. Simple parity is not enough.
- The degree of disproportion is not intended to be constant over the entire risk spectrum. The HSE defines the 'disproportion factor' as the ratio of sacrifice to risk reduction achieved. Under the ALARP concept, this disproportion factor would necessarily be greater as the risk approached the upper, intolerable risk level, and could be less for lower risks.
- The disproportion is intended, in part, to provide conservatism to compensate for any imprecision in the estimates of cost and benefit. See Section 2.8.1.
- Application of the ALARP principle does not relieve an organization of the responsibility for complying with recognized and generally accepted industry practices for risk controls.

2.8   UNCERTAINTY AND ITS IMPACT ON RISK DECISION MAKING

2.8.1   Sources of Uncertainty

These *Guidelines* have carefully used the terms *estimate* and *estimating* when speaking of modeling consequence and frequency and in determining the associated risk. The risk analyst and managers should retain an appropriate perspective of the potential error bounds associated with such estimates. Two important sources of uncertainty in risk estimates are knowledge uncertainty and modeling uncertainty. [HSE 2001]

Knowledge Uncertainty.   This can be viewed as the uncertainty associated with the assumptions and data input to risk estimates. For example, component failure rates may be based upon sparse statistics or subject to errors in the record keeping. Past failure rate data may not reflect the reliability and quality improvement of components as a result of technological advances. The assumption of atmospheric conditions used in dispersion modeling may not accurately reflect local conditions.   The potential impact of these uncertainties on the resulting risk estimates can be estimated by sensitivity analysis (see Section 2.8.2), which allows the identification and closer evaluation of particularly significant contributors to the uncertainty.

Modeling Uncertainty. This is associated with limitations in the validity of the mathematical calculations used to model the physical phenomena giving rise to the risks. For example, calculations of the flow rate through a relief valve may not adequately reflect the effect of multi-phase flow in the relief stream, or most dispersion models assume flat, open terrain while many process facilities are located on more complex topography.

The overall effects of these two types of uncertainty are to introduce, respectively, random and systematic errors into risk estimates. The awareness of such uncertainties prompts some risk analysts, as mentioned previously, to be more comfortable in making relative judgments based upon QRA results, rather than absolute judgments.   Certainly, these uncertainties must be considered when applying risk criteria to make judgments of risk tolerance.

2.8.2   Addressing Uncertainty

Risk criteria imply that clear boundaries exist between what is intolerable, what is tolerable, and what is broadly acceptable. Risk analysts and users of QRA results should not regard these criteria as being abrupt demarcations – risk criteria must be applied with a certain amount of common sense. For example, an analyst should not conclude that risk management efforts are sufficient just because a study shows the risk to be marginally below the maximum tolerable risk criteria.   Risk estimates are just that – estimates. While the estimated risk might be just below the maximum tolerable risk criteria, the true risk might be greater when considering the uncertainty of the analysis. Illusory perceptions of the accuracy and precision of the analysis should not be used to justify avoiding appropriate risk reduction efforts. Application of the ALARP concept helps ensure that systems that just barely satisfy risk criteria are still subject to management awareness and control in an effort to achieve further risk reductions.

Conversely, the uncertainty of the analysis should not be used as a justification for avoiding further risk reductions when the risk is above the maximum tolerable risk criteria. An assertion that, "Yes, we're above the line; but, considering the uncertainty, we could just as easily be below the line" is not a supportable justification for inaction.

Prudent consideration of the uncertainty inherent in QRA is an important aspect of risk decision making. While a comprehensive treatment of the topic of uncertainty is beyond the scope of these *Guidelines*, this chapter concludes with some basic observations on understanding and controlling uncertainty.

Uncertainty Analysis. Uncertainty analysis (or sensitivity analysis) allows QRA analysts to estimate the effects on the study results as a function of changes in data, models used, and/or assumptions made. While the possible combinations of sensitivity studies can be potentially overwhelming, experienced QRA analysts can usually focus in on the key variables and determine how they affect the overall risk estimates. When performing uncertainty analyses on QRA results that are to be compared to risk criteria, the rigor of the analysis is often related to the size of the margin between the risk estimates and the risk criteria (i.e., risk estimates that are clearly well below the risk criteria would receive less scrutiny). Common uncertainty analysis techniques are described in the *CPQRA Guidelines*. [CCPS 2000a]

Reducing Uncertainty. Many risk analysts believe that the key to reducing the uncertainty associated with the reproducibility of results is to standardize methodologies and databases. As noted previously, some companies are seeking to do this, and some regulators are requiring this. Organizations, such as CCPS, continue efforts to improve the accuracy of models and to broaden and improve the quality of databases. Such efforts should further reduce both knowledge and modeling uncertainties.

One component of the uncertainty of risk results not yet discussed, however, is the knowledge and skill of the analyst. QRA is a complex topic, and companies should require that practitioners be properly educated and experienced in its practice. Some companies have formalized programs for the development of risk analysts and quality control programs intended to ensure that risk studies receive appropriate internal peer review before being finalized.

*To measure is to know... If you can not
measure it, you can not improve it.*

Lord Kelvin (1824 - 1907)

3
LEARNING FROM REGULATORY PRECEDENTS

3.1 WHY STUDY RISK CRITERIA?

An organization seeking to establish risk criteria faces at least two challenges.
The first is to ensure appropriate technical rigor and practical applicability; in
other words, the risk criteria must be suitable for achieving the goal of
effectively guiding risk management decision making. The second challenge
is to ensure that the risk criteria can be viewed as being credible and equitable
– noting that the organization may be working in an environment where the
public, and perhaps employees, are skeptical of the levels of risk they perceive
being imposed on them.

As with any difficult task, learning from the experience of others who
have gone before is often helpful. With regard to the creation of risk criteria,
the broadest, and most fully documented, experience base is contained in the
literature describing the evolution of risk criteria, primarily for various
regulatory bodies. This chapter focuses on risk criteria precedents to aid the
reader in addressing both of the above challenges.

3.1.1 The Value of Risk Criteria Precedents

Section 2.6 introduced the concept of risk tolerance as a function of societal
values, noting, as the UK Health and Safety Executive (HSE) has suggested,
that 'the opinion of the public should underlie the evaluation of risk.' [HSE
1988] The regulatory risk criteria precedents described in this chapter were

often developed either with the opportunity for direct public input and scrutiny, or with the participation or under the review of publicly elected legislative bodies. In all cases, the risk criteria (and often, their bases) are publicly available (e.g., via the internet). Absent a public referendum, an open, transparent process for the development and application of risk criteria provides the greatest confidence that the risk criteria are reflective of societal values.

Unfortunately, an operating company does not have access to the mechanisms available to a regulatory body for seeking broad public engagement in the development of its risk criteria. However, by giving due consideration to the societal values reflected in existing regulatory precedents, a company can gain a measure of confidence that its efforts are consistent with those values that society has either explicitly or tacitly endorsed.

3.1.2   How This Chapter Is Organized

While quantitative risk assessment (QRA) was pioneered in the US, the UK and the Netherlands have led the consideration of how risk criteria could be established and used in evaluating QRA results in the chemical industry. Both of these countries continue to refine the development and application of risk-based evaluations of facility risks. For these reasons, this chapter focuses on risk criteria development in the UK and the Netherlands.

As discussed in greater detail below, the application of risk criteria within these two countries is somewhat different. Within the Netherlands, legislation requires a company to quantify risks and demonstrate compliance with certain criteria defined by the government. In the UK, no such formal requirement for risk quantification exists (at least for onshore operations). Rather, companies must demonstrate that risks are reduced as low as reasonably practicable (ALARP).[4] Companies within the UK often elect to use a quantified approach to meet this requirement for major hazard risks. Also, the UK HSE will usually apply a quantitative approach when providing its guidance to other parties (e.g., local planning authorities, public inquiries) on major hazards.

Sections 3.2 and 3.3 provide detailed descriptions of the precedents established within the two countries. Key milestones in the evolution of their risk criteria are highlighted.

---

[4] Specifically, the UK regulation requires that persons at work and those affected by work activities are safe *so far as is reasonably practicable* (SFAIRP). The HSE regards the demonstration of ALARP to satisfy the requirements for demonstration of SFAIRP. [HSE 2008b]   These *Guidelines* use the term ALARP when discussing the UK precedents.

Section 3.4 provides a comparative analysis of the two countries' risk criteria. It contrasts the somewhat different approaches used in developing and applying the risk criteria but concludes, in the final analysis, that the risk management results achieved in the two countries are quite similar.

Section 3.5 distills some relevant knowledge from the UK and Dutch precedents that a company may wish to consider when crafting its own company risk criteria. The reader seeking the 'bottom line' may wish to read Section 3.5 first and then return to the balance of the chapter for the detail.

The regulatory precedents documented in these *Guidelines* look at risk decision making from two complementary perspectives:

- The first perspective addresses the tolerability of the risk of facility operations (existing or proposed), seeking to determine if site- or process-specific risk management initiatives are either currently appropriate or require supplementation.
- The second perspective addresses risk decision making from what is typically termed the land use planning perspective. Here, the decisions to be made, typically by a regulatory or planning authority, focus specifically on whether or not to permit new off-site development in the vicinity of an existing hazardous operation or facility, assuming a given level of risk from the facility.

These *Guidelines* predominantly focus upon the first perspective, since most readers will be seeking guidance for crafting risk criteria to be applied by a company in evaluating the tolerability of risk. Land use planning precedents are discussed in Appendix B.

3.1.3   Other Precedents

Other countries have similarly pursued the development of risk criteria. Research indicates that most of these countries have drawn from the precedents established by either the UK or the Dutch models. Appendix B broadens the discussion of precedents in the context of a more global survey of these and other national, local, and industry risk criteria.

3.2   THE EVOLUTION OF RISK CRITERIA IN THE UK

The UK has more than four decades of experience in developing criteria for evaluating technological risks. The following sections, discussed chronologically, highlight some of the more significant milestones in the UK's continually evolving approach to risk management.

3.2.1   UK Atomic Energy Authority – 1967

F. R. Farmer first proposed an incident scale/frequency criterion for nuclear power reactors in 1967, as illustrated in Figure 3.1. [Farmer 1967] This figure is not a true societal risk F-N curve per se, since the frequency scale is not expressed as a cumulative frequency of 'N or greater' consequences. However, the Farmer curve quantitatively expressed the expectation that, as the magnitude of the incident consequences increases, the likelihood should decrease.

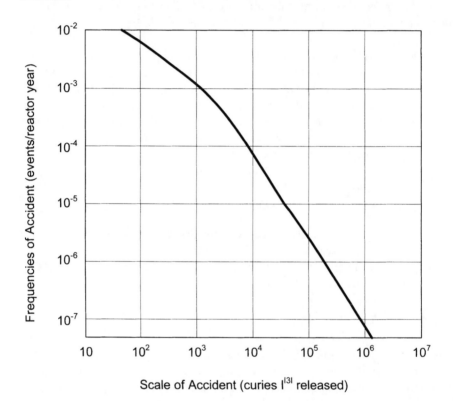

Scale of Accident (curies I[131] released)

FIGURE 3.1. Farmer Curve [Farmer 1967]

3.2.2   Health and Safety Commission/ Advisory Committee on Major Hazards – 1976

The Advisory Committee on Major Hazards (ACMH) was formed by the UK Health and Safety Commission (HSC) in 1974, largely in response to the vapor cloud explosion in Flixborough, UK. [Lees 2004] In its first report, the

ACMH addressed societal risk by suggesting that, for any one major non-nuclear plant, the maximum tolerable annual frequency for an undefined 'serious accident' might be 10^{-4}/year. [HSC 1976]

> "If... in a particular plant a serious accident was unlikely to occur more often than once in 10,000 years... this might perhaps be regarded as just on the borderline of acceptability, bearing in mind the known background of risks faced every day by the general public. [T]he distribution of casualties ... after a major accident ... may range from less than 10 to over 1,000... the upper part of this range is improbable as it requires the unlikely combination of many independent factors."

While the range of consequences was broadly defined, many experts interpret the report to imply that this 'serious accident' could involve 10 or more fatalities. In a subsequent publication, HSE quoted ACMH and noted: [HSE 1989a]

> "[ACMH]... which, for any **one** major non-nuclear plant, suggested 10^{-4} (one in ten thousand) as the maximum tolerable annual frequency for a "serious accident" – one undefined, but from the context, with the potential of killing 10 or more people, on or off site. For the 1500 or so Notified Installations in the UK, this implies a national frequency of such events of about 1 in 10 years. This is not inconsistent with UK experience."

As shown in Appendix B, the combination of 10 or more fatalities at an interval of 10,000 years has been used to establish the F-N anchor point of (10, 10^{-4})[5], which is still used in some risk criteria (e.g., Hong Kong; Sao Paulo, Brazil; and Victoria and New South Wales, Australia).

3.2.3   Royal Society – 1983

A Royal Society Study Group (RSSG), in 1983, defined concepts that the HSE would later integrate into its Tolerability of Risk (TOR) framework, proposed individual risk criteria that the HSE applies to this day, and discussed the concept of *as far as reasonably practicable* with respect to risk management. [RSSG 1983]   The RSSG proposed upper and lower bounds for risk to

---

[5] As a shorthand convenience, anchor points for F-N curves will be expressed in these *Guidelines* in the unitless form (N, F), where N is the number of fatalities on the X axis of the diagram and F is the corresponding frequency on the Y axis, in events/year, at which N or more fatalities are expected to occur. The entire F-N curve (assuming the curve has no inflection points; i.e., has a single slope along its entire length) will be expressed as (F, N, S), where S (a negative number) is the slope of the curve on a log-log graph.

workers of 10^{-3} fatality/year and 10^{-6} fatality/year, respectively. The HSE subsequently adopted the RSSG's proposals and substantiations, which are discussed in further detail below.

3.2.4   Health and Safety Executive – 1988

In its report, *The Tolerability of Risk from Nuclear Power Stations*, the HSE reaffirmed and elaborated upon the guidance contained in the RSSG report, extending the discussion to address societal risk. Even though this document focused primarily on the risk from nuclear reactors, it established the basis for discussion of risk in the UK for the next two decades and formalized the tenets upon which these *Guidelines* are premised. The TOR framework defined three basic tests that are applicable to the management of risk: [HSE 1988]

1. Is the given risk so great or the outcome so unacceptable that it must be refused altogether?
2. Is the risk so small, or has it been made so small, that no further precaution is necessary?
3. If the risk falls between these two extremes, has it been reduced to the lowest level that is reasonably practicable?

In addressing the third test, HSE detailed the concept of ALARP, noting that the 'higher or more unacceptable a risk is, the more, proportionately, that an employer is expected to spend to reduce it.' The TOR framework was depicted in Figure 2.11.

For individual risk, HSE suggested,

"Broadly, a risk of death of 1 in 1000 per annum is about the most that is ordinarily accepted under modern conditions for workers in the UK... and it seems reasonable to adopt it as the dividing line between what is just tolerable and what is intolerable."

A Comparison of Risks

In the TOR report, the HSE cites deep sea fishing (on UK-registered vessels) as posing an individual risk of 1.3×10^{-3} fatality/year (1990 data). This risk is comparable to the individual risk of 10^{-3} fatality/year reported by the US Bureau of Labor Statistics for fishing, hunting, and trapping in the US (2006 data). See Table 4.2. [BLS 2008]

As for the public, HSE recommended,

"[If the above is established for worker risk]... it seems to suggest that the maximum level that we should be prepared to tolerate for any individual member of the public from any large-scale industrial hazard should be not less than ten times lower, i.e., 1 in 10,000..."

Finally, having proposed upper bounds for tolerable risk, HSE proposed for 'broadly acceptable risk':

"This level might be taken to be 1 in a million (1 in 10^6) per annum bearing in mind the very small addition that this would involve to the ordinary risks of life... [this is] not altogether negligible... [but it] is a level of risk which, provided there is a benefit to be gained, and proper precautions are taken, does not worry us or cause us to alter our ordinary behavior in any way."

The HSE did not propose explicit societal risk criteria in this document, but did speculate on tolerable event frequencies, stating that:

"... a figure that might be accepted as tolerable for a considerable uncontrolled release [from a nuclear reactor] anywhere in the UK might be about 1 in 10,000 per annum..."

Note that the above frequency was a cumulative frequency for all nuclear reactors operating in the UK. HSE also projected that an event of this magnitude might result in 100 latent cancer deaths from radiation exposure.

3.2.5   Health and Safety Executive – 1989

In its publication, *Quantified Risk Assessment:Its Input To Decision Making*, the HSE summarized the application of QRA in practice at that time. [HSE 1989a]   In the discussion of societal risk, HSE used the nuclear incident consequence/frequency pair discussed above (100 latent cancer deaths, 10^{-4} fatality/year) as an F-N anchor point for a line with a -1 slope, noting that this slope 'would seem to reflect the least that we judge the public might require for larger N.'

3.2.6   Health and Safety Committee/Advisory Committee on Dangerous Substances – 1991

The report, *Major Hazard Aspects Of The Transport Of Dangerous Substances*, prepared by the Advisory Committee on Dangerous Substances (ACDS), looked at risks associated with road, rail, and marine transport where major incidents 'could affect non-employees or members of the public' in communities near dangerous goods transport routes and sea ports. [HSC 1991] The risks studied were those associated with releases of a variety of hazardous substances, notably liquefied petroleum gas (LPG), chlorine, ammonia, and gasoline.

The study adopted the TOR individual risk criteria and proposed societal risk criteria based upon the results of the Canvey Island study. The Canvey Island results were selected because these risks had been judged 'just tolerable' after considerable public and parliamentary scrutiny. In benchmarking the societal risk criteria against these publicly reviewed results, ACDS acknowledged the role for societal input in risk criteria promulgation, noting that "… the judgment on what is tolerable is not a scientific but a political matter."

The Canvey Island study results, after risk reduction improvements had been made, estimated that a major incident with 1,500 or more casualties (500 or more fatalities) would have an estimated frequency of once every 5,000 years. An F-N curve for maximum tolerable risk was plotted with the anchor point (500, 2×10^{-4}) and a slope of -1. ACDS selected this slope based upon the review of major incidents, asserting that the slope "reflects overall worldwide experience of events involving major installations." A parallel F-N curve for negligible risk was established three orders of magnitude lower on the frequency scale.

Canvey Island was a large complex containing multiple facilities (see sidebar) operated by a number of different companies. ACDS proposed applying the Canvey-based criterion for maximum tolerable risk at major ports around the UK; however, it recognized that the scale of operations at these different locations would vary considerably. To adjust for the scale of operation, ACDS proposed calculating a 'local scrutiny level' by scaling the Canvey-based risk criterion by

> **Facilities in the Scope of the First Canvey Island Study [Lees 2004]**
>
> - LNG or LPG terminals (2)
> - Petroleum and/or toxic liquids storage facilities (2)
> - Ammonium nitrate manufacturing plant, with associated anhydrous NH_3 storage (1)
> - Oil refineries, existing (2)
> - Oil refineries, proposed (2)
> - Large refinery tank farm with appreciable LPG and HF storage (1)
> - Local river, road, rail, and pipeline transport of hazardous materials

the volume of trade in tonnes at a particular port relative to the volume of trade at Canvey Island. The local scrutiny level would be informative in nature. For ports smaller in scale than Canvey Island, the scrutiny level would be below the risk criterion. Risks above the scrutiny level, but below the risk criterion, would still be tolerable, but would be flagged for further evaluation and possible reduction. The local scrutiny level was of no particular significance for ports larger in scale than Canvey Island, since the level would be above the risk criterion. Similarly, a national scrutiny level was proposed for evaluating transportation risks aggregated across the UK.

As these *Guidelines* will further explain, risk criteria should be commensurate with the scale of the activity to which they are intended to be applied. Criteria calibrated against a national risk perspective cannot logically be applied to decision making at the facility level. Similarly, criteria intended for evaluation of the risk from an entire enterprise should not be applied to decisions about individual risk controls at the scenario level.

3.2.7 Health and Safety Executive – 1992

HSE revised its *Tolerability of Risk* report in 1992, based upon public comments received on the original edition, noting: [HSE 1992]

> "… that final judgments about whether a given risk is tolerable, are not matters for experts alone, but for the people who have to bear the risks, and who are therefore entitled to be given the best possible technical advice about them."

HSE retained, for non-nuclear industrial plants and existing nuclear power stations, the individual risk criteria of 10^{-3} fatality/year and 10^{-4} fatality/year for maximum tolerable risks to workers and the public, respectively, and 10^{-6} fatality/year for broadly acceptable risk, "…with, of course, the ALARP principle applying to ensure that risk from most plants is in fact lower or much lower." However, HSE proposed a lowered 10^{-5} fatality/year criteria for public risk from new nuclear power stations:

> "… we propose to adopt a risk of 1 in 10^5 per annum as the benchmark for new nuclear power stations… this is, in the case of a new station, broadly achievable and measurable."

The above proposal notwithstanding, the individual risk criteria for the public remains at 10^{-4} fatality/year in the more recent HSE *Safety Assessment Principles for Nuclear Facilities*. [HSE 2006a] The report made no changes in the perspective on societal risk.

3.2.8 Health and Safety Executive – 2001

The HSE publication *Reducing Risks, Protecting People* was intended to "to set out an overall framework for decision making by HSE which would ensure consistency and coherence across the full range of risks falling within the scope of the Health and Safety at Work Act." [HSE 2001] This publication reflected considerable public input obtained from 150 responses to a solicitation for comments on an earlier discussion document that the HSE had released to the public. In *Reducing Risks, Protecting People*, the HSE noted that the crafting of risk criteria:

> "… will be the outcome of much deliberation and negotiation in the course of policy development, reflecting the value preferences of stakeholders and the practicability of possible solutions."

In the document, the HSE reinforced the TOR approach and reaffirmed the 10^{-3} fatality/year, 10^{-4} fatality/year, and 10^{-6} fatality/year individual risk criteria. However, the HSE noted that, in its experience, the individual risk limits for workers and the public 'rarely bite' – that is, the HSE suggested that societal risk issues commonly dominate the evaluation of risk tolerability.

Reducing Risks, Protecting People established a new F-N anchor point of $(50, 2\times10^{-4})$ for maximum tolerable public societal risk. This is the same frequency, but 1/10 of the N value, used by ACDS based upon the Canvey Island study results. This adjustment represented the HSE's scaling from the multi-facility Canvey Island context to risk criteria intended for a single major industrial activity, which HSE defined as:

> "…an industrial activity from which risk is assessed as a whole, such as all chemical manufacturing and storage units within the control of one company in one location or within a site boundary, a cross-country pipeline, or a railway along which dangerous goods are transported."

The HSE did not specify the slope of the societal risk F-N curve but, rather, referenced another publication for "a discussion of techniques available for extrapolating this criterion to other numbers of casualties and their frequency." This other publication, *Societal Risks* [Ball 1998], is not an official HSE publication – in fact, the title page conspicuously notes that the contents "do not necessarily reflect HSE policy." Nevertheless, *Societal Risks* goes into great detail regarding why an F-N curve slope of -1 could be regarded as appropriate.

Reducing Risks, Protecting People addressed a number of HSE risk assessment policy issues. Notably, HSE commented that:

> "We must also stress that these criteria are merely guidelines to be interpreted with commonsense and are not intended to be rigid benchmarks to be complied with in all circumstances."

Additionally, the HSE discussed the precautionary principle, which it applies when an apparent risk exists that cannot be adequately evaluated. In current practice, the HSE invokes the precautionary principle when "there is good reason to believe that harmful effects may occur..." but "the level of scientific uncertainty about the consequences or likelihood is such that risk cannot be assessed with sufficient confidence to inform decision-making." [HSE 2006b] The HSE's expectation is that, where threats of serious harm exist, lack of full scientific certainty shall not be used as a reason for postponing cost effective risk reduction measures.

3.2.9   Current Guidance to Industry and HSE Inspectors[6]

The HSE internet site contains a number of informative publications on risk management–related topics. One page on the HSE web site, *Hazardous Installations Directorate's (HID's) Approach to 'As Low As Reasonably Practicable' (ALARP) Decisions*: [HSE 2008a]

- Links, in a single HSE document, the 'proposal' of both the (50, 2×10^{-4}) anchor point and the -1 slope for the societal risk F-N curve, defining the maximum tolerable risk region for a single major industrial activity. It defines the curve for the broadly acceptable risk to be two orders of magnitude lower.
- Proposes that, where several sites contribute to the local societal risk, the intolerable risk region may be set an order of magnitude above that for a single major industrial activity (in effect, scaling upwards to larger facilities, perhaps like Canvey Island). This represents an example of risk scaling within a regulatory context.
- Provides insight on the HSE's perspective of what constitutes gross disproportion in an ALARP evaluation by establishing the expectation that the ratio of cost to benefit might be on the order of 1 or 2 where risks are close to being broadly acceptable. However, the ratio must be at least 10 at the intolerable risk boundary.[7]

---

[6] "Current" in this context means mid-2008, as the final draft of these *Guidelines* was being finished. Current events (e.g., the Buncefield terminal explosion [MIIB 2008]) may prompt changes in the guidance that HSE provides its inspectors.

[7] In making this cost benefit evaluation, HSE proposes using a benefit value of ~£1 million for each fatality prevented. Several other HSE guidance documents discuss the use of cost benefit analysis as a tool to assist in determining whether ALARP has been achieved. [HSE 2001 and HSE 2008b]

3.2.10 Other HSE Approaches To Addressing Societal Risk

The HSE has recognized that QRA requires a significant investment of time and resources, that it may not be practical for low risk situations, and that some organizations may not have the technical capability for conducting a rigorous QRA. In response, the HSE has sought to develop alternatives to conducting full QRAs, especially with regard to evaluating societal risk. One alternative pursued by the HSE is the risk integral previously mentioned in Section 2.4.2. [Hirst 2002] The HSE proposes that the risk integral be used as a screening tool by the HSE inspectors reviewing safety reports submitted under the Control of Major Accident Hazards (COMAH) regulation. The value calculated for the risk integral, when compared to relevant criteria, would indicate whether the risk was low enough to obviate the need for further study or, alternatively, whether more rigorous analysis was warranted.

3.3   THE EVOLUTION OF RISK CRITERIA IN THE NETHERLANDS

Section 2.1 mentioned the early application of risk criteria in the Netherlands when, in 1953, Parliamentary debate established a performance criterion (i.e., an individual risk criterion) of 10^{-6} fatality/year with respect to the adequacy of the sea dike systems. This section addresses the subsequent application of QRA and risk criteria to technological risks (primarily, facilities handling hazardous materials) in the Netherlands.

3.3.1   Groningen – 1978

The Province of Groningen, prompted by a series of major European process safety incidents in the 1970s, published both individual risk and societal risk criteria in 1978. [Groningen 1978] The individual risk criterion, set at 10^{-5} fatality/year, was equated to 1% of the total risk of <u>accidental</u> death to an average individual in the Dutch population. [Ball 1998] The societal risk criteria are shown in Figure 3.2. Ball reports that the criteria were to be applied to individual facilities handling dangerous goods. The N value was to be calculated as equivalent fatalities; in other words, one fatality (or serious injury) was equivalent to 10 injuries, 100 slight injuries, or 1,000 very slight injuries. The slope of the F-N curve was -2 for values of N greater than 1. Ball reports that this slope "was chosen on largely subjective grounds."

The Groningen risk criteria have an atypical, but not unique, characteristic – the abrupt termination of the curves at a certain value of N (several other criteria discussed in Appendix B, such as those for Hong Kong, incorporate this same feature). Criteria embodying this feature explicitly remove frequency from consideration and impose a deterministic, or consequence-based, approach to decision making for high values of N.

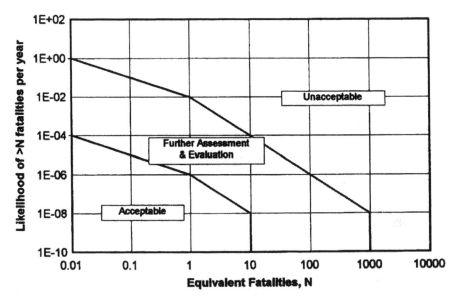

FIGURE 3.2. Groningen Societal Risk Criteria [Ball 1998]

3.3.2   Dutch Ministry for Housing, Spatial Planning, and the Environment – 1984

In the early 1980s, national government authorities in the Netherlands became increasingly aware of the potential consequences of a major industrial incident in a country as densely populated as the Netherlands and commissioned two major risk studies.

The Rijnmond study (mentioned in Section 2.1) was published in 1982. Recognizing the increasing use of LPG as a chemical feedstock and a fuel, including as a motor fuel, the Dutch later conducted an integrated risk study of the transport and use of LPG, issuing a 24-volume report in 1983. [TNO 1983] This latter study prompted a policy document, the *LPG Integral Memorandum*, which was prepared by the Dutch Ministry of Housing, Spatial Planning, and the Environment (VROM)[8] and accepted by the Dutch Parliament in 1984. [VROM 1984] This document served as the basis for Dutch external safety policy for the next two decades by establishing three main premises: [Voogd 2004]

Ale points out that the Dutch had a personal awareness of the hazards associated with the transport of liquefied flammable gases – of the more than 200 fatalities in the Spanish campsite disaster (Los Alfaques), 180 were Dutch. [Ale 2005b]

---

[8] In Dutch: Volkshuisvesting, Ruimtelijke Ordening en Milieu; VROM

1. The use of QRAs to evaluate risks
2. The requirement that both individual risk and societal risk be evaluated
3. The establishment of criteria for evaluating the tolerability of both individual risk and societal risk

---

Some notes on Dutch regulatory terminology:

In discussing the Dutch precedents, these *Guidelines* use the term *external safety*. In the Dutch regulatory context, external safety pertains to the "risks to the surrounding area in handling hazardous materials, in the use, storage, and transport of those materials." [VROM 2005]

The Dutch use the term As Low As Reasonably Achievable (ALARA), rather than ALARP, and these *Guidelines* will follow this precedent when discussing Dutch risk criteria.

In describing facilities handling hazardous materials, the Dutch use the following terms, consistent with the EU Seveso Directive (Seveso 2): [EU 1997]

- *establishment* shall mean the whole area under the control of an operator where dangerous substances are present in one or more installations, including common or related infrastructures or activities
- *installation* shall mean a technical unit within an establishment in which dangerous substances are produced, used, handled, or stored. It shall include all the equipment, structures, pipework, machinery, tools, private railway sidings, docks, unloading quays serving the installation, jetties, warehouses or similar structures, floating or otherwise, necessary for the operation of the installation

In larger establishments, docks, jetties, loading/unloading facilities, warehouses, and so forth, may be defined as separate installations.

Finally, in some regulatory contexts, the Dutch use the more general term *situation*. For example, a *new situation* could either refer to (a) provision of a new establishment, or the modification of an existing establishment, in the vicinity of an existing land use application or (b) provision of a new land use application, or the modification of an existing land use application, in the vicinity of an existing establishment.

The risk criterion for maximum tolerable public individual risk was set equal to the 10^{-6} fatality/year criterion previously established for the failure of seawall dikes. A societal risk F-N curve slope of -2 was selected to show a significant degree of risk aversion, based upon the precedent set by Groningen. However, the selected anchor point of $(10, 10^{-5})$ was an order of magnitude lower than that selected by Groningen. Broadly acceptable risks levels were set at 1% of the above values for both individual risk and societal risk, establishing ALARA regions between the broadly acceptable and maximum tolerable risk levels.

These risk criteria were applied to new activities falling under the scope of Article 5 (major hazard installations) of the 1982 Seveso Directive and for licensing and siting of LPG installations. [Ale 1991] Risk criteria for existing installations were to come later.

At about this same time, the Netherlands was actively supporting the development of computerized methods for conducting quantitative consequence and risk analyses. [Ale 1984]

3.3.3 Dutch Ministry for Housing, Spatial Planning, and the Environment – 1988/1989

The above individual risk and societal risk criteria were reaffirmed and formalized in the *Major Incident Risk Order* (BRZO)[9] [VROM 1988] which implemented the Seveso Directive in the Netherlands, and an associated interpretive document, *Premises for Risk Management.* [VROM 1989] Rather than retaining the linkage to the 1953 criteria for dike failure risks, the basis for the individual risk criteria of 10^{-6} fatality/year was derived by taking 1% of the 'natural death risk' for 10 to 14 year olds in the Netherlands. [Ale 1991] By either basis, the underlying premise was that the maximum tolerable individual risk should add only incrementally to the average risk of death for a person in the Netherlands.

The BRZO order also extended the application of the risk criteria to other risks, such as those associated with nuclear power and chronic exposure to toxic substances. In addition, coverage was extended to existing establishments. In recognition that existing establishments might have difficulty in achieving the same levels of risk as a new establishments, an individual risk criterion of 10^{-5} fatality/year was established for these establishments. No distinction was made between new and existing establishments for the societal risk criteria.

---

[9] In Dutch: Besluit Risico's Zware Ongevallen; BRZO

3.3.4   Dutch Parliament – 1993

In 1993, discussions in the Dutch Parliament resulted in changes to the risk criteria and how they were to be applied. [Tweede Kamer[10] 1993] First, the negligible risk levels were abolished for both individual risk and societal risk (with the intent that the ALARA principle would be applied to all levels of risk below the maximum tolerable level). Second, the maximum tolerable societal risk was made non-mandatory, with the intent that the local enforcing authority (e.g., the local permitting agency) should strive to maintain risks below the maximum tolerable societal risk – but could, in effect, waive the criteria given compelling justification.

"… the criterion for tolerability of societal risk was given a less strict character, in that exceeding the societal risk criterion can be tolerated by the authorities if sufficient arguments (e.g., economic) can be given for the specific situation involved." [Bottelberghs 2000]

Overall, the net effects of these changes were: [VROM 1995]

- The basic policies applied would still follow principles outlined in *Premises for Risk Management* [VROM 1989]
- The ALARA principle would remain the basis for local enforcing authorities to maintain risks below the maximum tolerable levels
- The notion of negligible risk was no longer part of the criteria
- Staying below the maximum tolerable individual risk was a mandatory requirement for new situations
- Existing situations not meeting the maximum tolerable individual risk would be prioritized:
- Risk receptors in the region between the 10^{-5} fatality/year and 10^{-6} fatality/year risk contours would be addressed as far as possible by 'technical' means
- Immediate actions would be required to address risk receptors inside the 10^{-5} fatality/year risk contour

3.3.5   Dutch Ministry of Transport, Public Works, and Water Management – 1996

In 1996, the individual risk criterion for dangerous goods transport was set at 10^{-6} fatality/year by the Ministry of Transport, Public Works, and Water Management (V&W).[11]   The corresponding societal risk criterion was

---

[10] The Tweede Kamer is the Second House of the Dutch Parliament
[11] In Dutch: Ministerie van Verkeer en Waterstaat

established one order of magnitude higher than the risk criterion for fixed establishments; in other words, $(10, 10^{-4}, -2)$. However, the F-N curve was intended to express the risk per kilometer of transport route. [Bekke 2006 and Bottelberghs 2000]

3.3.6   Dutch Ministry for Housing, Spatial Planning, and the Environment – 1999

An update to the BRZO order in 1999 introduced the distinction between 'vulnerable' objects and 'less vulnerable' objects. [VROM 1999] For both new and existing establishments, the individual risk criteria for less vulnerable objects (10^{-5} fatality/year) was set one order of magnitude higher than the risk criteria for vulnerable objects (10^{-6} fatality/year). The concept of 'less vulnerable' objects applied only to individual risk.

Examples of vulnerable objects would include: [Bottelberghs 2000]

- houses, apartment buildings, and other residential objects
- hospitals and other institutions for medical care
- schools and other education facilities
- objects of a high strategic value

Examples of less vulnerable objects would include:

- shops, department stores, and similar objects
- hotels, restaurants, and cafes
- commercial and industrial buildings
- office buildings
- recreational facilities

3.3.7   Dutch Ministry for Housing, Spatial Planning, and the Environment – 2004

The Netherlands made significant changes to its external safety framework in response to the Enschede [Voogd 2004] and Toulouse disasters. [Lees 2004] Modifications made to the risk criteria, and their basis for application, were documented by the publication of the External Safety (Establishments) Decree in 2004. [VROM 2004a] This Decree:

- Affirmed that individual risk was intended to be calculated as a 'location risk,' assuming that someone is exposed continuously, and without protection, at each location for which the risk is estimated.
- Affirmed that credit could be taken for factors such as occupancy patterns, indoor as contrasted with outdoor exposure, and other protections when calculating societal risk.

- Imposed statutory limit values on individual risk (location-based risk) for vulnerable objects for existing and new situations. Limit values must be achieved by a given date and not exceeded thereafter.
  - The limit value for new situations is 10^{-6} fatality/year.
  - For existing establishments with individual risk in excess of 10^{-6} fatality/year, the establishments may not be modified in any way that would increase that risk. The individual risk must be reduced to below 10^{-5} fatality/year by October 2007 and to below 10^{-6} fatality/year by January 2010.
- Imposed a target value on individual risk for less vulnerable objects. While numerically equal to the statutory limits applied to vulnerable objects, a target value is to be achieved so far as possible by a specified date and maintained so far as possible thereafter (i.e., it is not a hard limit). Target values may be exceeded only when compelling reasons for doing so are presented.
- Affirmed that the distinction between vulnerable and less vulnerable objects applies only to individual risk.
- Required relevant authorities to address the acceptability of societal risk when issuing permits to establishments or when adopting land use development plans. The Decree retained the original societal risk F-N curve [VROM 1984] as a non-mandatory societal risk benchmark.

A guidance document prepared by VROM, *Guidance on Justification of Societal Risk*,[12] established that the following populations would be included or excluded in the societal risk calculation: [VROM 2007]

- Employees of, and visitors at, the risk source are <u>excluded</u> (the visitor exclusion would extend to a situation such as visitors to a skating rink where the hazard is an ammonia refrigeration system).
- Employees of neighboring facilities (including other high hazard facilities) would normally be <u>included</u> when calculating societal risk (but, at the discretion of the relevant authorities, could be omitted based upon consideration of the emergency response preparedness of the neighboring facility).
- Persons on roadways, in train stations, and in public areas such as parks would be <u>excluded</u> from the societal risk calculations.

---

[12] In Dutch, *Handreiking Verantwoordingsplicht Groepsrisico*.

The societal risk calculation encompasses the population out to a distance from the source where the probability of fatality falls below 1%.

In summary, under current requirements, the only Dutch risk criterion for which compliance is mandatory is that for individual risk for vulnerable objects.

3.4   COMPARISON OF RISK CRITERIA IN THE UK AND THE NETHERLANDS

The current individual risk and societal risk criteria applicable in the UK and the Netherlands are summarized in Table 3.1 and Figure 3.3. The figure charts the F-N curves, which define the maximum tolerable societal risk criteria for the two countries, along with the criterion curve for broadly acceptable risk in the UK.

Different approaches and quantitative scales are evident in the contrasting criteria. This section seeks to put these differences into perspective, beginning with differences in the structure of the two systems, and then proceeding to a discussion of the significance of the orders-of-magnitude differences in the risk criteria.

The focus of this chapter is on precedents addressing the first context described in Section 3.1.2; in other words, the company's evaluation of the tolerability of the risk of facility operations (existing or proposed) to determine if site- or process-specific risk management initiatives are currently appropriate or if they require supplementation. The UK has separate criteria for use in this context, and for use in the land use planning context. In contrast, the Netherlands uses a common set of criteria for both contexts. The two countries' approaches to land use planning decision making are compared in Appendix B.

3.4.1   Individual Risk Criteria for Workers

The Netherlands does not have an individual risk criterion for workers. The UK is unique in addressing on-site individual risk in the regulatory context.

3.4.2   Distinction between Individual Risk Criteria for New and Existing Establishments

The UK does not explicitly reflect such a distinction in its current risk criteria. However, HSE guidance to its staff for making judgments of a facility's satisfaction of its ALARP responsibilities noted that the greater challenges faced by older establishments in retrofitting newer risk reduction technologies may be considered when evaluating what is reasonably practicable. [HSE 2008a and HSE 2008c]

TABLE 3.1. Comparison of UK and Dutch Individual Risk Criteria

Individual Risk Criteria, fatality/year	UK	Netherlands
Risk To 'Vulnerable Objects'		
Maximum tolerable risk to workers	10^{-3}	Dutch risk criteria do not address risks to workers
Maximum tolerable risk to the public	10^{-4}	New establishments: 10^{-6}
		Existing establishments: Beginning October 2007: 10^{-5} Beginning January 2010: 10^{-6}
Broadly acceptable (or negligible) risk to workers and public	10^{-6}	Risk criterion for negligible risk has been abandoned
Risk To 'Less Vulnerable Objects'		
Maximum tolerable risk to workers	UK does not make this distinction	The risk criteria above also apply to less vulnerable objects. However, the risk criteria are administered as target values rather than mandatory limits.
Maximum tolerable risk to the public		
Broadly acceptable (or negligible) risk to workers and public		

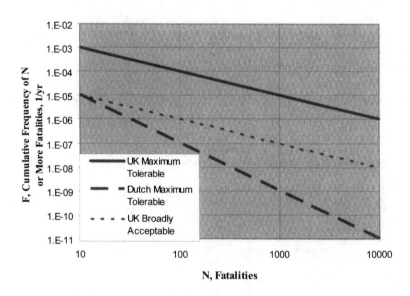

N, Fatalities

FIGURE 3.3. Comparison of UK and Dutch Societal Risk Criteria

3.4.3 Criteria for Broadly Acceptable (or Negligible) Risk

The Netherlands abandoned its negligible risk criteria for both individual risk and societal risk as part of a compromise that made the maximum tolerable societal risk criteria advisory rather than mandatory. In theory, then, the ALARA regions for individual risk and societal risk have no lower bounds. This could allow a more aggressive application of the ALARA principle. However, actual practice does not appear to achieve this intention.

3.4.4 The Distinction between Vulnerable and Less Vulnerable Objects

While the Netherlands makes the distinction between vulnerable and less vulnerable objects with regard to its individual risk criteria, the UK does not do so.[13] The significance of this distinction will be addressed in the discussion of the numerical disparities between the two countries' risk criteria presented in Section 3.4.6.

3.4.5 The Issue of F-N Curve Slope

One factor that makes the Dutch societal risk criteria so apparently conservative is the selection of a -2 slope for the F-N criterion curve. As noted previously, a slope of -1 (as used by the HSE) is regarded to be risk neutral, while more negative slopes are deemed to reflect increasing risk aversion (i.e., increasing concern with events that may result in large numbers of fatalities). Underlying this risk aversion is the premise that, for example, a single event resulting in 100 fatalities will prompt greater societal concern than 100 events with a single fatality each.

The discussion of F-N curves in Appendix A addresses the issue of curve slope and illustrates that even the 'risk neutral' F-N curve with a slope of -1 has a significant degree of unrecognized aversion to multi-fatality events associated with it.

3.4.6 A Perspective on the Numerical Disparities between the UK and Dutch Criteria

Sections 3.2 and 3.3 summarized the evolution of the risk criteria for the two countries. While not explicitly detailed, a fuller presentation of their history would show that each country periodically cited the other's work as precedents along the way. Why then, the orders-of-magnitude differences in the final risk criteria? Ale gives some insight into this when he contrasts the UK and Dutch approaches to risk management and risk criteria within the context of the two countries' respective legal systems. [Ale 2005b]

---

[13] The UK does, in effect, make this distinction in its land use planning criteria, as will be discussed in Appendix B.

Ale observes that the UK adversarial system, which is based upon English common law:

> "... leads to decisions in court where wrongful acts have to be compensated by paying damages... [which] in turn leads to consideration of what is a reasonable effort to avoid causing damage at a certain level of risk, when damages cannot be completely excluded... [which leads to the consideration of ALARP]."

In contrast, he proposes that, in the Dutch legal system, which is based upon Napoleonic law:

> "... the law defines what are unlawful or unjust acts and for each of these the penalty... is defined in the law... There is a constant history of verdicts... [that] should the government want to impose stricter rules, it should do so by law rather than through a court case..."

Ale goes on to assert that the 'gross disproportion' test required for ALARP in the UK allows authorities to demand the achievement of risk levels much lower than the specified risk criteria. He suggests that, in most industries, the actual risk levels achieved are at least an order of magnitude below the risk criteria. In contrast, in the Netherlands, the ALARA process for balancing the costs of risk reduction against the benefits derived does not require gross disproportion, thus "weighing risk reduction against costs is a much finer balancing act." [Ale 2005b]

Consequently, Ale's premise is that, in the UK, the risk criteria are the starting point, and ALARP considerations typically drive the risk further downwards. While, in the Netherlands, the risk criteria are commonly the end of the discussion in that "the courts invariably state that, should the government want more safety, it should put stricter levels in the law." [Ale 2005b] Other authors have observed that ALARA is often not a vital part of Dutch risk management efforts. For example, Bottelberghs has commented: [Bottelberghs 2000]

> "If safety measures would be extremely costly, this would not be considered reasonable in the sense of ALARA. As long as risk criteria are not exceeded, measures above 'ALARA' are not considered by the authorities."

Trbojevic is more direct, asserting: [Trbojevic 2004]

"If there is a lower risk situation than the norms require, the Authority can allow building up to the norm."

The UK's broadly acceptable risk criterion for individual risk is 10^{-6} fatality/year, equal to the Dutch criterion for maximum tolerable risk. However, differences exist in the individual risk calculations in the two countries that make direct quantitative comparisons difficult (see Section 3.5.6). Nonetheless, the authors previously cited leave the impression that two countries, in applying their risk management philosophies, may be approaching similar end results from different directions.

3.4.7   Difficulties in Satisfying the More Stringent Dutch Risk Criteria

History reveals problems in complying with the stringent Dutch criteria, especially with respect to societal risk.   Ale cites several examples: [Ale 2005a]

- In the early 1970s, the mayor of Dordrecht/Zwijndrecht petitioned Parliament to relax the societal risk criteria in order to allow large scale development of offices near railway stations. Here, the goal of facilitating mass transit commuting was in conflict with the risks associated with the adjacent large scale transport of hazardous materials.
- The second example involved the proposed expansion of the Schiphol Airport. The expansion was deemed essential to the Dutch economy, but would have resulted in a societal risk in excess of 10 times the total risk of the combined SEVESO sites in the Netherlands.
- Studies following the disastrous fireworks warehouse explosion in Enschede revealed a number of circumstances in which risk reduction measures should have been implemented previously in accordance with the Dutch regulations. These measures, which had been postponed because of cost considerations, included the relocation of 600 LPG fueling stations (for storage and sale of LPG as automotive fuel to consumers) that were too close to residences.

The Schiphol issue helped precipitate risk criteria revisions in 1993, which declared the criteria for maximum tolerable societal risk to be advisory rather than mandatory. As a consequence of this change, authorities could now tolerate exceeding the societal risk criteria if a compelling rationale (e.g., economic need) could be provided for the specific case. To offset the less stringent approach to societal risk, the concept of negligible risk was abandoned for both individual risk and societal risk. [Voogd 2004]

Thus, while the Dutch societal risk criteria are numerically more restrictive than the corresponding UK risk criteria, they are advisory rather than mandatory. They have been made advisory with the expressed intention of allowing authorities, with justification, to waive them. And, where the estimated risk is below the risk criteria, it is less likely (as compared to the UK) that ALARA considerations will drive the risk substantially lower.

Further revisions to the risk criteria framework, implemented under the 2004 *External Safety (Establishments) Decree*, [VROM 2004a] similarly liberalized the treatment of individual risk, relative to how such matters would be addressed in the UK. The concept of 'less vulnerable' objects establishes advisory rather than mandatory risk criteria for nearby exposures such as stores, hotels and restaurants, industrial buildings, office buildings, and recreational facilities. As with societal risk, these advisory risk criteria can be waived by authorities, given sufficient justification. The UK risk criteria framework makes no such distinction between different classifications of off-site populations (see Section 3.4.4).

Looking at the Dutch complement of risk criteria, only the individual risk criteria for vulnerable objects is mandatory in its application, and even then, more liberal risk criteria are allowed for existing vulnerable objects through 2009.

A joint report issued by the Advisory Councils of VROM and V&W makes some pointed observations about the implementation of Dutch risk management initiatives. These include: [VROM 2003]

- Risk generators are not provided with enough incentives to limit risks.
- In practice, the ALARA principle is often not effectively implemented, with some risk generators trying to maintain risk just below the risk criteria rather than trying to make further risk reductions.
- The concept of group risk (societal risk) was originally intended to apply to single stationary facilities, such as LPG fueling stations. However, the societal risk criteria are now applied to markedly different types and scales of facilities (such as large industrial facilities, railroad yards, and airports), complicating compliance with the criteria.
- Local authorities are often unwilling to enforce the external safety objectives.

- "There is a gap between the external safety level that the Dutch government wishes to offer its citizens and the effort it is willing to put into this. As a result, the aim of creating a living environment with limited external safety risks is often not achieved under the present system of standards."

VROM has drafted a document, *Guidance on the Accountability for Societal Risk*, to help all stakeholders, including local authorities, understand their roles in managing societal risk. [VROM 2004b] The degree of success in implementing the current risk criteria remains to be demonstrated and may be problematic in heavily industrialized, densely populated areas such as the Rotterdam-Rijnmond region.

Overall, the Dutch system is typified by numerically restrictive risk criteria, most of which have been loosened by making them advisory rather than mandatory, and by little application of the ALARA principle. In contrast the UK system entails less restrictive risk criteria, but greater emphasis on driving risk downward through the application of the ALARP principle. However, as Figure 3.4 suggests, despite their different starting points, the risk management systems in the UK and the Netherlands may be producing results that are not all that different.

3.5 LEARNING FROM AND APPLYING THE PRECEDENTS

Use of QRA and the associated risk criteria has evolved differently in the UK and the Netherlands, in response to country-specific drivers. Consequently, the precedents described above provide interesting parallels and contrasts. The balance of this chapter offers some observations that readers may wish to consider when using these precedents as a foundation for crafting risk criteria. Appendix B elaborates on this discussion in the broader context of a global sampling of risk criteria precedents.

3.5.1 Balancing Individual and Societal Risk Perspectives

Risk management programs can address individual risk, societal risk, or both. A tolerably low individual risk may not adequately represent the whole risk story if the individual is one of 20,000 in a crowded sports stadium. Similarly, the societal risk associated with a lightly used park area might be tolerable, while the individual risk for the resident caretaker could be significant. Both the UK and the Netherlands seek a balanced perspective through addressing both individual risk and societal risk in their criteria.

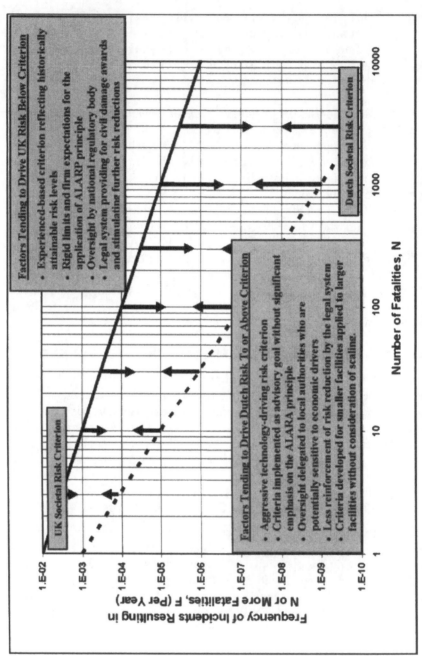

Figure 3.4. A Risk Performance Convergence

74

3.5.2 Balancing On-site and Off-site Risk Perspectives

The UK HSE is unique among the national regulatory bodies in quantitatively addressing risk to workers from catastrophic events, having promulgated explicit individual risk criteria for on-site populations.[14] While these *Guidelines* do not include data on company risk criteria, many companies have established individual risk criteria for workers, often giving heed to the HSE precedent.

While no regulatory body has implemented societal risk criteria for on-site populations, several companies have reported doing so. See Appendix C for the examples of how this might be done.

3.5.3 Experience-based Versus Technology-Driving Criteria

Two different approaches to developing risk criteria are evident in the precedents. Their nature and results should be kept in mind when interpreting and comparing risk criteria. These approaches can be characterized as:

- Experience-based – This approach, as applied in the UK, involves studying historical decisions about major hazards, particularly those subject to broad public discussion or scrutiny, and basing risk criteria on the established consensus. [Ball 1998]
- Technology-driving (Aspirational) – This approach, as applied in the Netherlands, involves establishing risk criteria that represent aggressive goals, whether or not experience indicates that such goals are currently attainable. The implication is that the goals will drive the development of the means for meeting the criteria.

Potential problems are associated with either approach. The experience-based approach (absent a counterbalance, such as the aggressive application of the ALARP principle, as in the UK) could freeze performance at past levels – even when improved performance is feasible (e.g., through the application of newer technologies). On the other hand, aspirational goals can reach too far too quickly, potentially creating a system in which failures to reach the aggressive goals degrade the credibility of the system.

3.5.4 Rigid Limits Versus Advisory Goals

Care should be taken to discriminate between those risk criteria that are intended to be rigidly enforced and those that are provided to be advisory in nature. Rigid limits must be practicably attainable to be credible. Advisory

---

[14] While the US Occupational Safety and Health Administration (OSHA) uses QRA to establish chemical exposure standards for workers, the focus is on lifetime cancer risks from chronic exposures.

goals, like the technology-stretching goals discussed above, should not stretch too far. Furthermore, since they are non-mandatory, advisory goals may not drive the risk reductions hoped for, or needed.

3.5.5   Application of the ALARP or ALARA Principle

The UK and Dutch precedents illustrate contrasting emphasis on seeking further risk reductions when the risk level is below the maximum tolerable criterion.   Where risk criteria are established in anticipation that the application of a continuing risk-reduction initiative, such as the ALARA or ALARP principle, will drive risk even lower, systems should be put in place to reinforce rigorous application of such initiatives.

As described above, the UK requires a gross disproportion between cost and benefit to justify no further risk reduction effort, while a more equal balance of cost and benefit might be deemed sufficient in the Netherlands. Where the ALARP or ALARA principle is applied, guidance should be provided with respect to the appropriate degree of disproportion expected between the resources required for risk reduction and the benefits derived.

3.5.6   Basis for Calculating Risks

To properly interpret and apply risk criteria, understanding the bases underlying the risk calculations is essential. The difference in the bases of the individual risk calculations in the Netherlands and in the UK illustrates this point:

- Individual risk in the Netherlands is calculated as a location risk. That is, the risk is calculated for a particular geographical location and is independent of the nature of the population, or even whether anyone is likely to be present at a particular location. The calculation assumes a receptor present 24 hours per day, 365 days per year, out of doors, with no protection (i.e., an *average unprotected person*). [Bottelberghs 2000]
- Individual risk in the UK is calculated for one or more *hypothetical individuals*. The concept of the hypothetical individual allows the risk estimate to reflect variations in exposure durations or patterns, diverse forms of protection available to different population groups, variations in individual characteristics that might affect susceptibility to harm (e.g., age, gender, abilities or disabilities), and so forth. Thus, multiple risk calculations may be required to determine the hypothetical individual at greatest risk at a particular location. [HSE 2001]

Valid decisions cannot be made if the protocol defining the risk calculation is not consistent with the basis underlying the risk criteria. The Netherlands has found it necessary, as part of a broader European Union (EU) effort, to standardize the modeling software, scenarios, and assumptions used to perform QRAs. This standardization was imposed to increase the consistency of the analyses and the decisions made based upon them, and to provide a basis for regulated parties to have greater confidence that they are in compliance with regulatory requirements. [DNV 2007]

3.5.7   Risk Criteria Scaling

Many companies seeking to develop risk criteria will have facilities of diverse sizes. For example, for some companies, a major facility might be an operation on the scale of an entire refinery, a large offshore complex, or an LNG manufacturing plant. QRA can be applied to operations that are larger than a major facility (e.g., an industrial complex comprising several refineries or several large processing plants). QRA can also be applied to operations that are smaller than a major facility (e.g., a single process unit within a refinery or a small or medium gas processing plant).

To provide the basis for rational risk management decision making, societal risk criteria are often established for a certain scale of operation and then adjusted to reflect the scale of the system being evaluated. Failure to appropriately scale the criteria could result in higher risks than warranted being tolerated for a small facility, or the inordinate expenditure of risk mitigation resources to reduce the risk for a very large facility.

This chapter discussed scaling in the application of the UK criteria, and referenced difficulties in the application of the Dutch criteria as a consequence of not scaling. The issue of societal risk criteria scaling will be addressed further in Chapter 4.

3.5.8   Application of New Risk Criteria to Old Establishments

The precedents discussed in this chapter and in Appendix B demonstrate various approaches to the issue of whether individual risk criteria should be applied with equal rigor to existing establishments and to new establishments where, presumably, more advanced risk management technologies or approaches might be practicably available.

The approaches described for the UK and the Netherlands, with respect to addressing existing establishments when creating or implementing individual risk criteria, were:

- Rigidly apply the risk criteria to existing establishments, but allow a grace period for compliance – Netherlands (vulnerable objects)
- Make the risk criteria advisory (non-mandatory) for existing establishments – Netherlands (less vulnerable objects)

- Acknowledge that it may not be reasonably practicable to retroactively apply to existing facilities the same controls required to reduce risks ALARP in a new facility (i.e., reducing risk ALARP in an existing facility may leave a residual level of risk greater than could be achieved in a new facility) – UK

Application of new risk criteria to old establishments may need to be considered and explicitly addressed when crafting and implementing new risk criteria, especially if the risk criteria are intended to prompt aggressive reductions in existing risk levels.

3.5.9   Risk Criteria for Transportation Risk

As Appendix B will show, many countries have found transportation risk to be a difficult issue to address. This is the case primarily with respect to societal risk, since the risk aggregated over the entire length of the transport route can be far greater than the risk of a single fixed facility. As described in Section 3.3.5, the Netherlands has established explicit risk criteria for transportation risk. The UK has made no such distinction; aggregating transportation activities with fixed facilities in its definition of a 'single major industrial activity' (see Section 3.2.8).

Where there is much to risk,
there is much to consider.

Anonymous

4
CONSIDERATIONS IN DEVELOPING
RISK CRITERIA

This chapter addresses how organizations may utilize the information in previous chapters to develop risk criteria that are:

- Consistent with societal expectations
- Technically sound
- Scalable, reflecting variations in company operations
- Understandable to users and consumers of risk information alike
- Effective in helping businesses evaluate and manage the risk of their operations
- Consistently applicable worldwide, for multinational companies

Sections 4.1 through 4.11 describe the step-by-step process, illustrated in Figure 4.1, for developing, validating, and maintaining risk criteria. Sections 4.12 through 4.14 describe some additional considerations that organizations may wish to keep in mind when developing and implementing risk criteria.

The development of risk criteria is an important task that can have a significant impact on how an organization conducts its business and should not be undertaken without the endorsement and support of upper-level management. Hendershot describes a number of important considerations, including seeking to involve all potentially impacted parts of the organization in risk criteria development. Organizations may elect to form a multi-discipline team to implement the guidance provided in this chapter. Such a team might include: [Hendershot 1995]

- Safety, health, and environmental personnel
- Research, development, and engineering personnel
- Operations personnel
- Business and upper management personnel
- Risk analysis professionals

Other expertise that might be represented include: legal, public relations, technical, maintenance, or other plant or corporate groups.

4.1   DETERMINE THE NEED TO ESTABLISH RISK CRITERIA

The most basic question for the organization seeking to develop risk criteria is, "Why do we need them?" The organization must have a sound understanding of how it is going to use the risk criteria – what questions it seeks to answer with the aid of the risk criteria – before it proceeds to develop them.

Hendershot lists four principal reasons for performing a quantitative risk analysis (QRA). [Hendershot 1995]

- Identify the major contributors to facility risk, in order to more effectively target risk-reduction activities and resources
- Compare design alternatives, with respect to their impact on facility risk
- Identify the predominant contributors to residual risk for attention in ongoing risk management efforts (e.g., to avoid making changes that might increase the risk)
- Make absolute judgments about the tolerability of the risk level

The first three objectives in the above list involve comparisons of relative levels of risk and can be accomplished without risk criteria. The organization should undertake the task of developing risk criteria only if it intends to accomplish the fourth objective.[15]

---

[15] A fifth objective could be added to the list – conduct of QRA to satisfy regulatory requirements. However, in this circumstance, the risk criteria may be provided by the regulatory agency requiring the QRA.

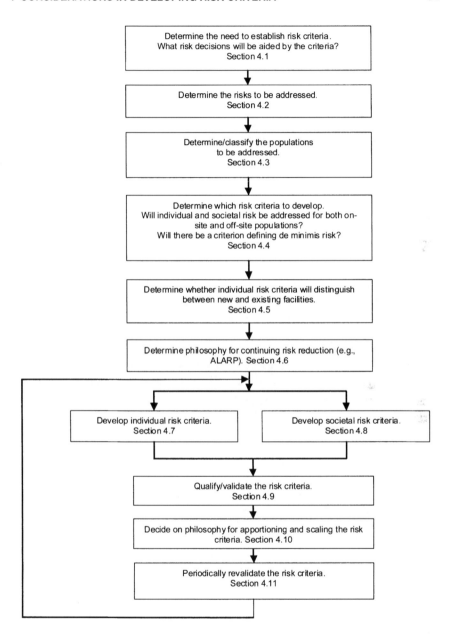

FIGURE 4.1. Risk Criteria Development Flowchart

4.2   DETERMINE THE RISKS TO BE ADDRESSED

As noted in Chapter 1, the focus of these *Guidelines* is on risks of human health effects associated with the hazards of process operations. Specifically the *Guidelines* focus on:

- Consequences associated with episodic events (i.e., events of limited duration rather than events such as chronic releases)
- Acute health effects (i.e., adverse health effects with severe symptoms developing rapidly) leading to fatalities

Other conventional hazards associated with industrial operations (i.e., more traditional worker safety issues) are typically not included in assessments using quantitative risk criteria as described in these *Guidelines*, but are addressed primarily with qualitative criteria.

Some organizations may seek to address a diverse spectrum of risk types in addition to fatalities from process risks; for example, they might choose to quantify environmental damage or damage to company reputation. The Swiss precedent described in Appendix B is an example of such an attempt.   The Groningen societal risk criteria discussed in Section 3.1.1 illustrated another variation: by incorporating the concept of equivalent fatality, it included various degrees of non-fatal injuries into the consequence figure. [Groningen 1978] Organizations attempting to quantify a diverse spectrum of risks face the challenge of understanding the relative significance of the various types of risk in relation to one another.

An organization intending to develop risk criteria must clearly identify the types of risk it seeks to address.   The balance of this chapter focuses exclusively on the risk of fatalities from episodic events.

4.3   DETERMINE/CLASSIFY THE POPULATION TO BE ADDRESSED

The risk criteria precedents described in these *Guidelines* illustrate a variety of approaches to classifying at-risk populations that may be addressed by risk criteria.   Some of the more significant distinctions are summarized below. Organizations may not choose to make all of these distinctions or to address all identified populations types (e.g., an organization may elect to focus only on off-site risks).   In any case, the risk criteria must be defined consistent with the populations to which they will be applied.

4.3.1   On-site populations

On-site populations are typically employees, contractors, and site visitors.   As noted in Section 2.6.3, and exemplified by some of the precedents described in these *Guidelines*, society typically tolerates higher risk when the risk is voluntarily assumed and when the person exposed to the risk is perceived to have the awareness, skills, and resources to help mitigate the risk. Employees, who enter the site to seek the benefits of employment, who are trained on process hazards, and who are provided personal protective equipment, are generally regarded to share in the responsibility for being cognizant of, and seeking to manage, the risks associated with their employment.

On-site contractors, who should be educated and prepared with respect to the hazards they are exposed to, are typically grouped with plant employees with respect to risk criteria applications. However, one distinction that can be made involves contractors on construction projects isolated from process units. Such projects typically have separate access gates, and these workers, who are not typically working near process hazards, may not be as well prepared as on-site contractors. Such project contractors, who do not receive similar training or do not have access to personal protective equipment or facilities to shelter in place as do on-site workers, might be considered part of the off-site population.

While the case for site visitors is less straightforward, they too are typically included in the on-site category as they are typically either educated on the hazards they may encounter or are escorted by knowledgeable site employees.

4.3.2   Off-site populations

Off-site populations might include (but are not limited to):

- Residential populations
- Occupants of hospitals and long-term care facilities
- Students in schools
- Inmates of correctional institutions
- Shoppers in commercial settings
- Persons outdoors in parks or other recreational facilities (even if the facilities are on company property)
- Employees in other industrial facilities

Off-site populations are assumed to be less prepared and less capable of protecting themselves from process risks. And, since they often do not experience the direct benefits of the facility (e.g., employment), they are regarded as involuntary risk receptors. For these reasons, risk criteria for off-site populations are often lower than for on-site populations.

As described in Chapter 3, the Netherlands makes the distinction between *vulnerable* and *less vulnerable* off-site populations. Similarly, as shown in Appendix B, some land use planning criteria characterize different levels of 'vulnerability' for different classes of off-site populations. In such systems, the underlying risk criteria might allow higher risk exposures to workers in adjacent industrial plants when contrasted to the risk allowed for, say, residential populations. Where such neighbors are part of the same industrial complex and have received training in responding to hazardous situations as part of a coordinated emergency response plan, treating them similarly to on-site populations may be appropriate.

Chapter 3 and Appendix B illustrate that, where the distinction is made between worker and public risk, the public (off-site) risk criteria are often one full order of magnitude lower than the on-site risk criteria (often, for both individual and societal risk).

4.4   DETERMINE WHICH RISK CRITERIA TO DEVELOP

The precedents in these *Guidelines* illustrate various perspectives on the number and types of risk criteria to apply. In order to identify the risk criteria applicable to its needs, an organization must address some basic questions about its risk management philosophy.

4.4.1   Will the Risk Criteria Address Both Individual and Societal Risk?

Individual risk, as the term suggests, provides a perspective on facility risk from an individual's point of view, whether a worker or a member of the public. In contrast, societal risk provides a perspective on the risk to the company, especially from potentially catastrophic events. As important as every individual is, the reality is that the potential impact on the company increases for multiple fatality events, in contrast with individual fatalities. While the organization has the option of determining which types of risk it will address, industry is becoming increasingly aware of the benefits of a balanced risk perspective that considers both individual and societal risk (albeit, either could be addressed qualitatively).

4.4.2   Will the Risk Criteria Define a De Minimis Level of Risk?

Section 2.7 introduced the UK Health and Safety Executive (HSE) Tolerance of Risk (TOR) framework, which was illustrated in Figure 2.11.   The framework broadly defined three risk regions:

- Unacceptable
- Tolerable, but warranting consideration for reduction
- Broadly acceptable, not requiring further reduction

An upper risk criterion (e.g., *Higher Risk* or *Maximum Tolerable Risk*) and a lower risk criterion (e.g., *Lower Risk* or *Broadly Acceptable Risk*) define the boundaries (thresholds) between the three regions. Between these criteria is an As Low As Reasonably Practicable (ALARP) region where further efforts would be expected to seek reasonable and appropriate risk reductions. The TOR diagram in Figure 4.2 illustrates these boundaries. [HSE 1992]

Some companies and regulatory authorities (such as the Netherlands, as described in Chapter 3) elect to omit the lower risk criterion – communicating that no level of risk is so low as to warrant ceasing consideration of efforts to further reduce the risk. This extends the ALARP region down the diagram, without limit, as illustrated in Figure 4.3.[16]

The election of which option to follow (i.e., a three-tier risk criteria model, with a de minimis level, or a two-tier risk criteria model, without a de minimis level) is a philosophical decision that each organization must make: does the organization wish to establish a continuing improvement emphasis on risk reduction or, alternatively, indicate that a point exists beyond which consideration of further risk reductions is not obligatory? In reality, some organizations using the three-tier risk criteria model still encourage employees to identify opportunistic risk reductions, even in the broadly acceptable region, if these can be achieved with small incremental effort or cost.

The concepts outlined in Figures 4.2 and 4.3 are applicable to both individual and societal risk. Note that the selection of either model for individual risk does not constrain the organization to apply the same model to societal risk (or vice versa).

4.5   DETERMINE WHETHER RISK CRITERIA WILL DISTINGUISH BETWEEN NEW AND EXISTING FACILITIES

Organizations that are interested in crafting new risk criteria, or realigning existing risk criteria to new risk perspectives, may have some existing facilities built to standards and technologies less advanced than those currently available.[17] While upgrades may have occurred over the years, some facilities might still have difficulty meeting more stringent risk criteria that are aligned with performance expectations for newer facilities designed to implement more advanced risk controls. New industry standards, technology, and ALARP drive risk lower on new projects where changes are easier to make prior to construction.

---

[16] Note that, to aid communication of the comparison between the two TOR models, Figure 4.2 [HSE 1992] has been modified to create Figure 4.3. Only the two-criteria model, as shown in Figure 4.2, is described in the referenced HSE publication.

[17] In this context, 'facility' could be either an entire site under the control of a single organization or a single production unit on a site.

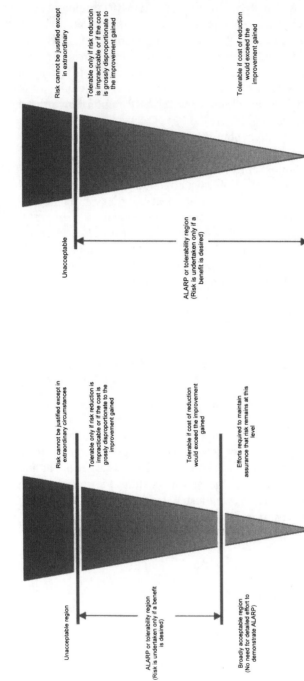

FIGURE 4.2. TOR Diagram with De Minimis Level
[HSE 1992]

FIGURE 4.3. TOR Diagram without De Minimis Level

Some risk criteria systems provide allowances for such situations – in effect, lowering risk management expectations for existing facilities, at least for a period of time. Section 3.5.8 describes options for addressing this issue, with respect to individual risk criteria, that were distilled from the regulatory precedents.

4.6   DETERMINE PHILOSOPHY FOR CONTINUING RISK REDUCTION

Some application of the ALARP principle, as discussed in Sections 2.7 and 4.4, is inherent in the implementation, and is essential for success, of the TOR framework. Fundamentally, the ALARP principle drives continuing risk reduction. Without ALARP, the maximum tolerable risk criterion may be perceived as the final goal of risk reduction efforts; in other words, efforts could cease once the risk level is reduced to less than the criterion.

Risk criteria should be crafted with an appreciation of what ALARP means to the organization. The contrast between the UK and Dutch approaches to risk criteria and ALARP, described in Chapter 3, illustrates the importance of this point. Will the organization establish experience-based risk criteria and drive risk downward with rigorous application of the ALARP principle (as in the UK)? Or will the organization establish technology-driving (or aspirational) risk criteria and place relatively less reliance on ALARP (as in the Netherlands)? The appropriate choice may be linked to the organization's culture (e.g., does the organization have a continuous improvement philosophy or is it more accustomed to rule-based approaches?). Such cultural issues are beyond the scope of these *Guidelines*.

4.7   DEVELOP INDIVIDUAL RISK CRITERIA

While Section 2.4.1 describes various ways of expressing individual risk criteria, these *Guidelines* assume that most companies will use the more common approach of expressing the risk criteria as a fatality rate (i.e., fatality/year).

The decisions made in Sections 4.3 through 4.6 will define what constitutes a full complement of individual risk criteria for the organization. For example, this could be any or all of the following:

* Maximum tolerable risk for the public, new facilities
* Maximum tolerable risk for the public, existing facilities
* Maximum tolerable risk for workers, new facilities
* Maximum tolerable risk for workers, existing facilities
* De minimis risk for workers and the public

4.8   DEVELOP SOCIETAL RISK CRITERIA

These *Guidelines* assume that most companies establishing societal risk criteria will elect to use a F-N curve to display/implement these risk criteria.

As discussed previously, the two pieces of information defining a basic F-N curve are the anchor point and the slope of the curve. The slope of the curve is selected to reflect the organization's degree of risk aversion. A slope of -1 is regarded as being 'risk neutral' (although, as described in Appendix A, the cumulative frequency aspect of the F-N curve actually embodies an inherent degree of risk aversion for a -1 slope). Some organizations elect to reflect their risk aversion by using a higher slope or by including an inflection point in the F-N curve and increasing the slope above a certain value of N. For example, a slope of -1 up to the inflection point and -2 beyond it would indicate that the company is 'risk neutral' for smaller values of N, while seeking to be more conservative (i.e., more risk averse) at higher values (see Appendix C, Section C.2).

The decisions made in Sections 4.3 through 4.6 will define what constitutes a full complement of societal risk criteria for the organization.

4.9   QUALIFY/VALIDATE THE RISK CRITERIA

Any company that elects to develop and apply risk criteria should take steps to validate that the risk criteria are fair, credible, and defensible. This section describes several approaches that may be effective in qualifying or validating specific risk criteria.

4.9.1   Relating Risk Criteria to Commonly Tolerated Levels of Risk

People tolerate a certain amount of risk in their lives daily. Some of this risk is consciously accepted in return for the benefits it provides (e.g., air travel that allows us to reach a destination more quickly). Other risks may be so commonplace that we consciously acknowledge them only infrequently (e.g., slipping and falling in the shower) or they may be so obscure that the average person may not even know they exist (e.g., the health risks of botulism in some canned goods).

Some companies and regulatory bodies have sought to benchmark worker risk criteria against historic levels of work-related risk that are presumably tolerated by the general public. For example, as described in Chapter 3, the UK HSE established its maximum tolerable individual risk criterion for workers at 1 fatality in 1,000 years, based upon the observation that this is about the highest risk observed (and tolerated) in the highest risk industries, such as deep sea fishing. [HSE 2001] Table 4.1 illustrates some commonly accepted levels of worker risks, based upon US fatality statistics for the year 2006. [BLS 2008]

TABLE 4.1. Individual Risk of Accidental Work-Related Death in US, 2006, Fatality/Year [BLS 2008]

All private industry	4.3×10^{-5}
Fishing, hunting, and trapping	9.8×10^{-4}
Aircraft pilots and flight engineers	9.0×10^{-4}
Crop production	3.4×10^{-4}
Coal Mining	3.0×10^{-4}
Taxi drivers	1.9×10^{-4}
Transportation	1.7×10^{-4}
Construction	1.3×10^{-4}
Gasoline station	9.8×10^{-5}
Protective services (police and firemen)	9.7×10^{-5}
Wood products manufacturing	6.7×10^{-5}
Federal government	2.8×10^{-5}
Chemical manufacturing	2.7×10^{-5}
Plastics and rubber products manufacturing	1.8×10^{-5}
Professional, scientific, and technical services	9.1×10^{-6}
Hospitals	4.6×10^{-6}
Education, training, libraries	2.8×10^{-6}

With respect to off-site individual risk, criteria are often benchmarked against the average rate of non-work-related accidental death within the general population. However, to address the issue of involuntary versus voluntary risk, those using this approach often set the risk criteria at some fraction of the commonly tolerable risk level. The premise, then, is that the additional risk described by the risk criteria will be only a small fraction of the risk willingly assumed by the public. The Netherlands used this approach when establishing its public individual risk criterion by taking 1% of the 'natural death risk' of a 10- to 14-year-old Dutch resident. [Ale 1991] Table 4.2 illustrates some typical accidental death risks, based upon US fatality statistics for the year 2003. [NSC 2007a and 2007b]

Statistical data sufficient to create historical F-N curves are, in comparison, much more difficult to come by. Regulatory bodies have more commonly based societal risk criteria on past QRA studies whose results, after considerable public or legislative review, were deemed to represent a tolerable

TABLE 4.2. Individual Risk of Accidental Death in US, 2003, Fatality/Year
[NSC 2007a and 2007b]

Deaths due to unintentional (accidental) injuries	3.8×10^{-4}
Fatal occupational injuries, unintentional deaths	3.4×10^{-5}
Falls, all types	5.9×10^{-5}
Car occupant	5.4×10^{-5}
Complications of medical and surgical care	9.8×10^{-6}
Uncontrolled fire in building or structure	9.5×10^{-6}
Air and space transport accidents	2.6×10^{-6}
Firearms discharge, accidental	2.5×10^{-6}
Accidental suffocation and strangulation in bed	1.7×10^{-6}
Drowning and submersion while in or falling into bathtub	1.1×10^{-6}
Contact with venomous animals and plants	3.2×10^{-7}
Lightning	1.6×10^{-7}
Bitten or struck by dog	1.1×10^{-7}
Explosion and rupture of pressurized devices	1.0×10^{-7}

risk in consideration of the benefits gained from the activity under study. An example would be the UK HSE societal risk criteria that are ultimately traceable to the results of the Canvey Island study. [HSE 2001]

A company might also develop its societal risk criteria based upon consideration of the impact of multi-fatality events on the reputation and sustainability of the company. Appendix C provides such an example.

4.9.2   Relating Risk Criteria to Societal Values Reflected in Regulatory Precedents

Some companies believe that the determination of tolerable levels of risk is largely the responsibility of governmental or regulatory bodies (or, possibly, industry associations), and only then with the participation of the potentially affected public and workforce. Chapter 3 and Appendix B describe a number of examples illustrating how governmental agencies sought public and/or legislative review of proposed risk criteria during their creation or update. For example, the HSE describes a very open and collaborative process whereby considerable public input and commentary was sought as risk criteria were proposed, refined, and made official. [HSE 2001]   In addition, as these *Guidelines* go to press, the HSE is soliciting public input on a number of matters related to risk assessment issues.   Similarly, Dutch risk criteria have evolved with significant involvement of the Dutch Parliament. The UK and the Netherlands are notable, but not the only, examples of such engagement.

Unfortunately, most companies do not have such opportunities for public input when they are crafting their risk criteria. And, absent public involvement, such companies are subject to potentially negative reactions from those whom they are seeking to protect – especially members of the public who may perceive that risks are involuntarily imposed upon them.

Consequently, many companies have risk criteria that are based upon, or benchmarked against, established regulatory precedents. Drawing upon regulatory precedent allows the company to assert that the risk criteria are consistent with risk levels reflective of public (societal) values. Appendix C provides examples of how corporate risk criteria might be developed and validated against regulatory precedents.

4.10   DECIDE ON A PHILOSOPHY FOR APPORTIONING AND SCALING THE RISK CRITERIA

Consider an organization establishing and implementing a criterion for maximum tolerable societal risk. A risk analyst seeking to properly apply this criterion must understand the basis underlying the criterion. For example, is it intended to be applied to:

- The total risk of all operations across the company?
- The total risk of all operations at a particular site?
- The total risk of all operations within a process unit at a site?

What if the company has two operating sites that differ markedly in factors such as their physical size, surrounding population, inherent hazards of materials handled, value of production, and so forth? Should these two disparate sites be given the same societal risk criteria?

Similarly, consider an organization establishing and implementing a criterion for maximum tolerable individual risk for members of the public in the vicinity of a company site. Of the total amount of risk allowed for a single individual, how much should be permitted to come from:

- A particular process unit at a site?
- A specific operation within that unit?
- A single incident scenario?

Organizations may evaluate and seek to manage risk at all levels within the company, as illustrated in Figure 4.4. Unfortunately, some organizations have established risk criteria without providing clear guidance on how, or at what level, the risk criteria were intended to be applied. For example, comparing the individual risk of a single process operation – or of a single incident scenario – to a risk criterion intended for a whole facility ignores the fact that other facility operations will also contribute risk.

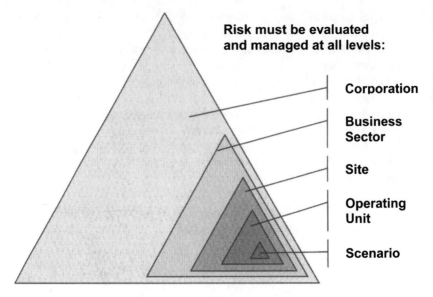

FIGURE 4.4. Levels at which Risk Must be Managed

This section of the chapter deals with two related concepts in the application of risk criteria: apportionment and scaling.

Apportionment. Apportionment refers to taking a risk criterion (which could be viewed as a risk allotment) and subdividing it among a number of risk sources. The concept of risk apportionment can apply to both individual and societal risk, but at different levels in the enterprise. For example, a company might envision a certain amount of societal risk that it is willing to assume at the corporate level and then apportion this total amount of risk to its various business sectors, which then apportion it to the various plant sites and perhaps to individual operating units. An example of this process is presented in Appendix C and discussed in Section 4.10.1.

Individual risk is local in its application, in that the individual is impacted by the facility the individual is on or adjacent to. In contrast to the treatment of societal risk, thinking in terms of a corporate-level individual risk criterion that is somehow subdivided among various business or operating units would be illogical. Individual risk can be apportioned, but only at the site level and below. Thus, the site individual risk criterion might be apportioned to the various operating units on the site and even down to the scenario level, as discussed in Section 4.10.2.

Figure 4.5 illustrates, for a hypothetical organizational structure, the levels in the organization over which societal and individual risk could be apportioned. In this example, societal risk criteria are promulgated at the

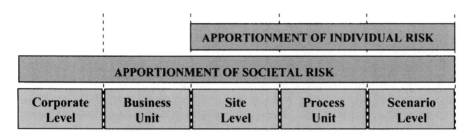

FIGURE 4.5. Levels over which Risk Criteria May be Apportioned

corporate level and apportioned down through the organization to the scenario level. Individual risk criteria are promulgated at the site level and apportioned down to the scenario level. Figure 4.5 is only intended to be illustrative. Organizations will have differing organizational structures.

Scaling. Scaling refers to adjusting the relative size of the risk portions, at a given level, based upon factors such as scale of operation. For example, as the societal risk allotment is apportioned to the various sites within a business sector, the most equitable apportionment may call for sites to receive differing shares of the allotment, as discussed further in Section 4.10.1.

Most organizations would agree that individual risk criteria do not scale; that is, the maximum risk tolerated for an individual is constant at the various sites across the corporation and is not a function of some site-specific scaling factor, such as the magnitude of the operation.

4.10.1 Apportioning and Scaling Societal Risk Criteria

Section C.1 in Appendix C describes one approach to apportioning and scaling societal risk criteria across a corporation to illustrate the issues that might come into play when promulgating risk criteria at this level.

In the example, the company first established an amount of societal risk at the corporate level that it regarded to be 'unsustainable' (i.e., an aggregate risk level that the company must not exceed). This maximum level of risk was then apportioned to the various business sectors within the company. Each of the company's business sectors were of comparable scale, in terms of financial turnover and number of employees; consequently, the company decided to apportion the corporate risk criteria equally among the business sectors (i.e., the company did not need to apply any scaling considerations when determining the risk allotment for the various business sectors).

The company had greater variability among the business sectors, with some business sectors consisting of a single operating site and some having 100 or more sites. By apportioning the risk allotment equally between the business sectors, and then further apportioning the risk allotment equally within each business sector, the company created a situation in which the societal risk allotment applicable to some large sites was 100 times greater than the risk allotment applicable to some small sites. The company regarded this factor appropriate considering the relative scales of operation at the various sites.

Improper scaling/apportionment of risk criteria introduces the potential for poor risk decisions, leading to either inadequately managed risk on one hand or over-managed risk (potentially with wasted effort and resources) on the other. A success measure for proper risk criteria scaling might be that comparable situations yield consistent decisions at different sites.

Societal risk criteria scaling is a potential issue at all levels of an enterprise. A variety of different factors could be considered when scaling the apportionment of the risk criteria:

- Physical size (e.g., plot plan area)
- Size of workforce
- Throughput
- Revenue
- Capital investment
- Inventory of hazardous materials
- Community impacts
- Equipment count
- Business criticality (including infrastructure)

Clearly, the degree of direct correlation to the hazard potential associated with site operations varies for the above factors; that any single hazard-relevant variable would provide a comprehensive perspective is unlikely. Unfortunately, little precedent is available for guidance; the topic of risk scaling provides an opportunity for future research efforts to develop a technical approach to solving the problem.

The issue of risk scaling is reminiscent of the approach proposed by the Advisory Committee on Dangerous Substances (ACDS) in the UK for monitoring dangerous goods transportation risks at ports (see Section 3.2.6). ACDS proposed risk scrutiny levels that would be established for each port based upon throughput; however, an upper risk criterion that was the same at all ports was also proposed. [HSC 1991]

Proper controls should be established to prevent inappropriate application of risk criteria scaling/apportionment. For example, dividing an existing site between two different business sectors to create two financially independent entities might, under some risk apportionment approaches, lead to an unwarranted doubling of risk allotment.

While this section focuses principally on scaling and apportioning societal risk criteria between sites, similar considerations exist, for example, between process units within sites. Regardless of the approach used, the organization should consider reserving some portion of the risk allowance for future expansion (e.g., the addition of new sites at the business unit level or the addition of new process units at the site level).

4.10.2 Apportioning Risk Criteria to Individual Scenarios

Ultimately, risk management at an operating site is the product of myriad individual decisions – some made ad hoc, but many made in more formalized contexts, such as process hazard analyses (PHA), layer of protection analyses (LOPA), and management of change activities. Some decisions can be made in a qualitative fashion, but others warrant quantitative criteria. The discussion of apportioning risk criteria must address a reasonable basis for assessing the tolerability of the risk contribution of individual event scenarios to the aggregated risk of site operations.

As stated in Section 2.4.1, the estimation of individual risk in a risk study is typically based on the summation of risk contributions from many event scenarios that could potentially cause a fatality to an individual; in other words, the sum of all risks due to fires, explosions, toxic chemical exposures, dropped objects/external impact, and so forth to which the worker is exposed. These risks may be identified in many different types of hazard and risk analyses. For facility personnel, some risks are localized to a specific process unit where the individual works. Other risks may be attributable to incidents in other process units at the same facility or associated with non-process activities. Some risks are related to start-up, maintenance, and shutdown activities

> *Scenario:* An unplanned event or incident sequence that results in a loss event and its associated impacts, including the success or failure of safeguards involved in the incident sequence.
>
> *HEP Guidelines*, 3rd edition [CCPS 2008]

during which worker exposure to the potential hazard and occupancy in the affected area may be high, but the duration of exposure may be brief. Other risks are related to random failures of equipment leading to process deviation from intended operation.

As companies implement new controls to reduce overall risk, determining the risk reduction requirement for the new controls, based on consideration of the total risk, can be challenging. For simplicity and uniformity, companies could consider establishing individual hazardous event risk criteria based on a conservative estimate of the total number of fatality-related events that a typical worker or member of the public might be exposed to.

Determining a reasonable basis for such estimates is essential. At issue, then, is how many hazardous events, collectively across the whole facility, might be identified during analytical exercises such as PHAs and risk analyses – counting only those events for which fatalities are projected as a credible outcome, absent risk controls. Keep in mind that, consistent with the basis described in Section 4.2, these *Guidelines* focus on risk criteria addressing fatality risk associated with process operations.

The total number of identified hazardous events across a facility will depend on a number of factors, including:

- The nature, scale, and complexity of operations at the facility. The total number of potentially fatal events is likely to be higher for more complex operations or for larger facilities.
- The level of detail at which the hazard identification (HAZID) is conducted. For example, in a high-level facility safety report, the description of a hazardous event may be of the form, "vapor cloud explosion on distillation unit." However, a hazard and operability analysis (HAZOP) will usually examine events at a much finer level of detail, so that a typical event description may be of the form, "vapor cloud explosion on distillation unit due to overfill of column V320." The total number of hazardous events will increase as the level of detail in the HAZID exercise is increased.
- The range of hazards addressed. In some cases, the risk assessment process will cover only a limited range of hazards. For example, when assessing the required reliability of a safety instrumented system (SIS), only events that can be mitigated by the addition of the SIS may be addressed (i.e., other potential failure mechanisms, such as loss of containment events, are excluded from consideration).[18]

Depending upon the scale and complexity of the operation and the level of detail of the HAZID, the total number of identified hazardous events involving a potential fatality across an entire facility could range from fewer than 10 to more than 1,000. When apportioning individual risk criteria to individual scenarios, a key consideration is what proportion of these events could impact a given individual.

An individual's vulnerability to an event depends on the location of the individual relative to the event and the hazard range of the event. For most identified hazardous events, an individual must be close to the event for a fatality to occur, although some hazards (e.g., toxic gas) may endanger people at a greater distance.

---

[18] Refer to CCPS *Guidelines for Safe and Reliable Instrumented Protective Systems* [CCPS 2007b] for information on safety instrumented systems and the assessment of required reliability for such systems.

For workers, individual risk is normally dominated by events in the unit in which the individual is working at the time.[19]   For example, a large, complex process unit may have as many as 100 hazardous events, identified by a LOPA or HAZOP, that could potentially cause a fatality of a given worker. This number of potential events would suggest that, for the purposes of HAZOP, LOPA, and other analyses at a similar level of detail, setting the target individual event criterion two orders of magnitude below the maximum individual risk criterion for workers would be reasonable. A number of regulatory precedents discussed in these *Guidelines* establish a maximum individual risk criterion of 10^{-3} fatality/year for workers. Based on the above example, the individual risk allotment for a single scenario with the potential for worker fatalities could be set at 10^{-5} fatality/year.

Members of the public will generally be farther from the process than facility workers, and will usually be potentially impacted by fewer events associated with a particular process unit. In addition, public risk is less likely to be dominated by the risk from a single unit; in other words, a given member of the public is more likely to be exposed to the risk from multiple units. For risk analyses at the level of LOPA or HAZOP, a particular member of the community may potentially face as many as 100 scenarios associated with a large facility, such as a refinery or a large manufacturing facility, where the scenarios are dominated by flammability hazards. If the company's risk criteria apply to QRAs of a large manufacturing site handling toxic materials, the number of scenarios might be appreciably larger because the affected distances for toxic releases are typically greater.

A number of regulatory precedents discussed in these *Guidelines* establish a maximum individual risk criterion of 10^{-4} fatality/year for members of the public. Continuing the above example, this would suggest that, for the purposes of HAZOP, LOPA, and other analyses of a similar level of detail, the individual risk allotment for a single scenario with the potential to cause off-site (public) fatalities might be on the order of 10^{-6} fatality/year.

In a similar manner, a company could attempt to apportion risk from its societal risk criteria to individual scenarios. This approach is more challenging given the larger number of potential scenarios that could result in one or more fatalities and the need to apportion the risk at the different levels of consequences illustrated on a societal risk F-N curve.   One method used by some companies is to allocate a portion (e.g., 1% or less) of the overall societal risk criterion to each scenario (i.e., establish individual scenario targets two orders of magnitude below the overall societal risk criteria for 1%).

-----

[19] Some employees, such as maintenance workers, may be exposed to the risks of multiple process units. However, unless widely disparate risk levels exist in these various units, the analysis presented above is still valid.   The total number of potential fatality scenarios that such a worker is exposed to will be a weighted average based upon the fractions of the work week spent in each unit.

When using either of these methods for apportioning risk to individual scenarios, some companies, when all the studied scenarios for the facility have been evaluated, sum the risks from the individual scenarios to ensure that the total risk does not exceed the overall individual or societal risk criteria – which may be possible for facilities with hundreds of scenarios.

The examples in this section are hypothetical. Each organization should, based upon its own experience, determine the apportionment factors that are appropriate to the nature and scale of its operations, the degree of resolution in its HAZID processes, and the range of hazards addressed.

4.10.3 Calibrating the System Over Time

An ultimate measure of the adequacy of the risk apportionment would be whether or not comparable risk decisions are being made in comparable situations. While performing comprehensive comparisons of all risk decisions made across a corporation is not feasible, spot checking of high-risk scenarios should be considered. Where disparate risk decisions are evident, what are the underlying cause(s)? Are these decisions attributable to differences in the analytical approach, or are they driven by the risk criteria used?

Hendershot discusses the need for credibility in risk criteria. Risk criteria that are so liberal that they allow facilities with obvious risk problems to slip through are not credible. Conversely, risk criteria should be reevaluated if they consistently fail facilities having state-of-the-art designs and systems meeting or exceeding industry standards. [Hendershot 1995]

4.11 PERIODICALLY REVALIDATE THE RISK CRITERIA

Organizations should regard their risk criteria as being dynamic and should periodically revalidate them – organizational needs and societal risk perceptions may change with time. Catastrophic events (such as Bhopal, Seveso, and Piper Alpha – all described in [Less 2004]) can have global impacts on public attitudes. Other changes can occur more gradually, as public perceptions shift (see Section 2.6.3). Finally, the organization itself may decide that enhanced management of process risks is warranted and decide to tighten its risk criteria.

An Example of Apportioning Individual Risk Criteria for HAZOP and LOPA

Some teams conducting studies such as a HAZOP or LOPA may find it more convenient to have a criterion expressed in terms of the tolerable frequency of an event that is likely to result in a fatality. One approach suggested for crafting such a criterion involves first apportioning the individual risk on a scenario basis as described above, then back calculating the tolerable frequency assuming a typical probability of experiencing a fatality for a typical scenario.

For any particular event, the probability of fatality for a given individual can be determined in terms of the probability that the given individual is present when the event occurs (occupancy) and, if present, the probability that the given individual will be fatally injured (vulnerability).

For on-site risk, an individual's occupancy will depend on working hours. For example, a worker on a 40 hour week will have an occupancy fraction of approximately 0.24 (40/168).

An individual's vulnerability to an event depends on the location of the individual relative to the event and the hazard range of the event. Some events may be associated with a very high probability of fatality for the given worker, whereas, for other events, the probability may be low. An estimate of 50% vulnerability (0.5 probability of fatality) might be appropriate across the full range of events on a unit.

The example in the main text for on-site risk showed a per-scenario individual risk criterion of 10^{-5} fatality/year (a site individual risk criterion of 10^{-3} fatality/year apportioned to 100 scenarios). Based upon the assumptions above (0.24 occupancy fraction and 0.5 fatality probability for the worker, if present), the corresponding tolerable frequency for the scenario would be:

Frequency, event/year $= (10^{-5}$ fatality/year) / (0.24 x 0.5, fatality/event)
$\approx 10^{-4}$ event/year

For off-site (public) risk, assuming an occupancy fraction of 1.0 might be more appropriate (reflecting an individual who is home most of the time). Because of greater separation distances, the average vulnerability to an event for members of the public would be expected be lower than the corresponding worker vulnerability – an assumed fatality probability of 0.1 might be more appropriate in these circumstances. Building upon the example in the main text for risk to an off-site individual, the corresponding tolerable frequency for the scenario would be:

Frequency, event/year $= (10^{-6}$ fatality/year) / (1.0 x 0.1, fatality/event)
$= 10^{-5}$ event/year

For these examples, the resulting criteria would be:

Criteria	Worker	Public
Maximum Individual Risk - All hazardous events	10^{-3} fatality/year	10^{-4} fatality/year
Maximum Individual Risk - Each hazardous event	10^{-5} fatality/year	10^{-6} fatality/year
Maximum Frequency - Each hazardous event	10^{-4} event/year	10^{-5} event/year

An Example of Apportioning Societal Risk Criteria in Risk Matrices

As Section 2.4.2 pointed out, societal risk can be depicted by means other than F-N curves, and some of these may be amenable for apportioning societal risk down to the scenario level. A risk matrix is one way that this apportionment could be accomplished. Assume, for example, that the consequence scale of a risk matrix is expressed in ranges of the number of fatalities potentially resulting from a single scenario. If one such range covered 3-10 fatalities, a value of 6 could be assumed as the approximate logarithmic mid-point of the range, and a value for F could be determined from the site F-N curve for a value of N = 6. This cumulative frequency (for all scenarios) could then be divided by the number of scenarios estimated, as described in Section 4.10.2, to determine a maximum tolerable frequency. By repeating this calculation for the midpoint of each range of consequences, consequence-frequency pairs could be determined to create a demarcation in the risk matrix corresponding to the maximum tolerable societal risk for a single scenario.

4.12 MATCHING THE RISK CRITERIA WITH THE SYSTEM AND WITH THE RISK ESTIMATION METHODOLOGY

To promote valid risk decisions, the organization should take care to provide a consistent approach to defining 1) the risk criteria, 2) the scope of the particular study (i.e., defining the system being studied), and 3) the risk assessment methodology.

4.12.1 Matching the Scope of the Study to the Basis for the Risk Criteria

To illustrate one issue addressed by the above statement, consider Figure 4.6, which displays societal risk F-N curves from the WASH-1400 study, the seminal US nuclear power industry risk study conducted in 1975. [NRC 1975] These curves suggest the total US societal risk from a number of natural causes and man-made activities. Plotting the risk of a single chemical plant process unit, or even the risk of the entire site, against these curves and drawing a favorable conclusion would be inappropriate (*but some risk analysts have done this!*). The size of the system under study and the context under which the risk criterion is intended to be applied must be consistent. For example, if the risk criterion is intended to be applied at the site level, repeatedly judging the risk of individual process units against the whole of this risk criterion would be inappropriate; doing so would overdraw the risk budget for the site.

4.12.2 Providing Consistency Between the QRA Methodology and the Risk Criteria

Hendershot observes that risk criteria are analogous to product specifications. When a specification is established, a defined test method is usually developed for generating the numbers to be compared with the specification. [Hendershot 1995] Valid conclusions can be drawn only if the correct test method is applied. Similarly, the results of a QRA can be highly dependent upon the analytical models selected, the assumptions made, and the data used. To make valid risk decisions across an entire enterprise, provisions must be made to promote consistency in studies made by different analysts in different locations.

Section 2.3.4 discussed company standardization of modeling tools, data, assumptions, and analyst qualifications. As noted in Section 3.5.6, the Netherlands has found it necessary to standardize and mandate a common software tool and modeling assumptions to promote consistency of QRAs. Similar approaches are being implemented in other new European Union land use planning guidance.

4.12.3 Clearly Defining the Basis for the Risk Criteria

Research of regulatory precedents revealed a common problem – the discussion of risk criteria without establishing the exact basis for their application. For example, does the individual risk criteria discussed apply to the maximum individual risk, or average individual risk? For consistent application, the basis for the risk criteria should be clearly defined and communicated.

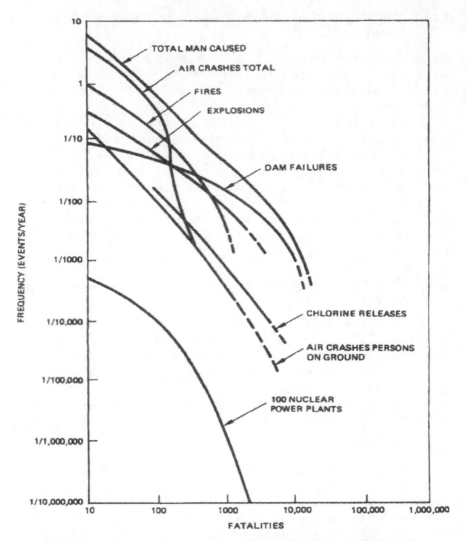

FIGURE 4.6. Societal Risk Curves from the WASH-1400 Study [NRC 1975]

4.13 RISK CRITERIA IN THE MULTINATIONAL CONTEXT

As Appendix B indicates, few regulatory risk criteria precedents exist within the US. However, this is not the case within Europe, South America, and the Asia/Pacific region, where countries are increasingly promulgating mandatory risk criteria. Multinational companies operating both in the US and abroad should consider the diverse regulatory contexts they may encounter when establishing their own risk criteria.

Some countries have established very restrictive risk criteria that may not be workable in a global context. Consequently, companies may not elect to base their own risk criteria on the most restrictive regulatory precedent that they might encounter. Instead, they may choose to establish risk criteria that make sense to them, with the proviso that they will comply locally with more restrictive risk criteria, where such exist.

Although some cultures may be more risk tolerant, many companies choose not to apply, in other countries, risk criteria that are less protective than they would apply at home. Also, risk perceptions are rapidly evolving, especially in developing countries. Companies should remain alert to new regulatory initiatives that might impact portions of their global operations. (See Section 4.11.)

Finally, the estimated societal risk for a facility may be more likely to change with time for a facility located in an area with a rapidly growing economy and population. While this is a risk analysis issue, rather than a risk criteria issue, companies should anticipate the need for more frequent updates of risk studies in such regions.

4.14   CONSIDERATION OF SHORT DURATION/ HIGHER RISK ACTIVITIES

The history of major incidents highlights a frequent problem in risk analyses – the failure to properly consider the elevated risks of certain short-duration activities, such as start-ups after maintenance turn-arounds. While this is a problem associated more with risk analysis than with risk criteria, its significance warrants a brief mention here.

Focusing on turn-around start-ups as the example, two principal concerns exist: 1) the nature of the activities increases the likelihood (i.e., expected frequency) of a significant event, such as a large hydrocarbon leak and 2) a significantly higher number of workers may be at risk within the unit during such activities. The first concern increases both individual and societal risk, and the latter concern further increases the societal risk – for the duration of the start-up.

*The best way to predict the future
is to invent it.*

Alan Kay (1940 -)

5
FUTURE TOPICS

The topic of risk criteria has evolved, and continues to evolve. These *Guidelines* address a brief history of risk assessment, summarize relevant precedents where regulatory bodies and other organizations have promulgated risk criteria, and outline an approach that a company could use in crafting and calibrating its own risk criteria. This chapter documents opportunities for future work in risk decision making, divided into two general categories:

- Fostering the Application of Risk-Based Decision Making. These topics address ways in which risk professionals can better educate stakeholders on the concept and value of risk-based decision making and the need for risk criteria within and beyond their organization.
- Enhancing Risk Assessment Technology. All technologies improve over time. These topics stem from the risk assessment professionals' natural desire to sharpen their tools and methodologies needed to assess risk and define and apply risk criteria.

5.1 FOSTERING THE APPLICATION OF RISK-BASED DECISION MAKING

Enhancing awareness of risk concepts. Risk professionals should promote risk awareness – both outside and within the industry. For example, these *Guidelines* point out that the highest consequence incident scenarios are not necessarily the highest risk incident scenarios.   This concept may be counterintuitive to many. Risk professionals should establish a dialog in order to educate company business and technical leadership. Leaders, in turn, set the tone for the acceptance and implementation of the risk-based approach in decision processes. They are also better positioned to advocate the value of these concepts to external stakeholders.

Providing educational materials. These *Guidelines* are a useful resource to risk professionals seeking to educate peers and decision-makers alike – along with other Center for Chemical Process Safety (CCPS) publications, such as *Guidelines for Chemical Process Quantitative Risk Analysis, Second Edition* [CCPS 2000a] and *Tools for Making Acute Risk Decisions with Chemical Process Applications.* [CCPS 1995] As CCPS continues to produce valuable guidance documents for industry, it may look to the field of risk-based decision making for additional relevant topics to address. Included in this category would be training materials that risk professionals could use to train company executives and other decision-makers.

5.2   ENHANCING RISK ASSESSMENT TECHNOLOGY

Calibrating qualitative risk matrices against quantitative risk criteria. The use of qualitative risk matrices in process safety management applications continues to increase – particularly in concert with efforts such as process hazard analysis (PHA), layer of protection analysis (LOPA), and risk-based inspection.   These *Guidelines* provide perspective on the importance of ensuring consistency between qualitative decision tools, such as risk matrices, and the quantitative risk criteria that should underlie them. The discussion of risk criteria apportionment in Section 4.10 contained a proposed means of assigning risk criteria allotments to individual scenarios. However, additional research is needed on the topics of risk criteria apportionment and scaling.

Comparing and calibrating risk concepts across diverse industries. While the *Guidelines* subcommittee members shared the basic concept of risk as a function of consequences and likelihood, an important lesson learned by the subcommittee was that different process industries view risk issues from different perspectives.   Differences such as the predominant hazards (e.g., toxic versus flammable), nature and scale of operations (e.g., small batch versus large continuous), and organizational structure (e.g., US-based versus multinational) can produce unique challenges with respect to the evaluation and management of process risk. These *Guidelines* sought to identify and record consensus. Opportunities remain for continued dialog and additional work targeted to specific industry segments.

Documenting risk criteria and matching to methodology. A common theme throughout these *Guidelines* is the interrelationship of the risk assessment methodology and the risk criteria against which the results are assessed. For example, using a maximum individual risk criterion to evaluate the results of an average individual risk analysis will not promote sound risk-based decision making. As companies develop their own risk criteria, they should keep this important concept in mind. However, companies will not be the only organizations crafting risk criteria. Regulators and industry associations promulgating risk criteria should ensure that the assumptions and the basis for their application are clearly and completely documented. In researching regulatory precedents, the *Guidelines* committee noted that a number of regulatory criteria were not effectively documented.

Addressing transportation risk criteria. These *Guidelines* note a few regulatory precedents for risk criteria developed to specifically address transportation risk assessments. However, the main focus of the *Guidelines* is on risks associated with the operation of fixed facilities. Additional research is required to address the unique risk decision needs and challenges posed by transportation risk assessments. CCPS has developed new *Guidelines for Chemical Transportation Safety, Security, and Risk Management*, 2nd Edition. [CCPS 2008]

Providing guidance on continuous risk reduction concepts. As these *Guidelines* note, continuous risk reduction plays an important role in the management of process risks. The opportunity for the development of additional guidance on this topic exists, including guidance on the application of cost-benefit considerations.

Calibration of diverse types of risks. These *Guidelines* focus on risks associated with episodic events causing acute health effects leading to fatality. Businesses face a spectrum of risk types which, if unmanaged, could jeopardize their sustainability. These include, but are not limited to, challenges to security, environmental performance, business continuity, company reputation, and so forth. As company leadership seeks to balance and manage multiple types of risk, the challenge becomes how to judge the equivalency of such diverse risks. [Farquharson 2007] Additional research on this topic is warranted.

Keeping apace of regulatory developments. A factor that complicated the research for these *Guidelines* was the fact that regulatory precedents continue to appear, evolve, and mature. A mechanism or venue for staying abreast of future developments and keeping the community of risk professionals informed of new or revised regulatory and industry association risk criteria precedents is needed. Furthermore, some references are only available in the language of the source nation. A centralized database of translations into one or more commonly used languages would be of value.

Appendix A
Understanding and Using F-N Diagrams

Section 2.4.2 introduced the F-N diagram as a means of presenting information about societal risks. F-N diagrams may be used to depict at least three different types of information: [Ball 1998]

- The historical record of incidents
- The results of a quantitative risk analysis (QRA)
- Criteria for judging the tolerability of risk

This appendix first illustrates the development of an F-N curve to show historical data or QRA results, and then concludes with a discussion of the use of F-N curves to communicate risk criteria. The significance of the slope of the F-N curve, when used to display risk criteria, has been mentioned

Performing QRAs

Readers wishing more detailed background on performing QRAs are directed to the CCPS publication, *Guidelines for Chemical Process Quantitative Risk Analysis.* [CCPS 2000a]

elsewhere in these *Guidelines*. This appendix provides further perspective on this topic.

A.1 CONSTRUCTION OF AN F-N CURVE TO DISPLAY DATA

The following discussion addresses the use of F-N curves to display consequence/ frequency data; specifically, an example of an F-N curve for communicating QRA results is described. The procedure for constructing the curve is the same whether the analyst is dealing with actual consequences (i.e., the historical record of incidents) or projected consequences (i.e., QRA results).

A.1.1   The Mathematics of the F-N Curve

During the analysis to estimate societal risk, various hypothetical events will have been assessed. Each of these events will have a predicted frequency of occurrence, f, and a predicted number of persons harmed, N. Consistent with the focus of these *Guidelines*, N in this discussion will be the predicted number of fatalities associated with the event. The associated frequency of occurrence will be expressed in events per year.

Many analysts believe that societal risk results are most effectively presented in graphical form. Such plots are normally log-log plots with the x-axis representing the consequences and the y-axis representing the frequency of occurrence. Log-log scales are used because the range of values for f and N can span multiple orders of magnitude.

Societal risks may be plotted in either of two fashions:

- Non-cumulative frequency basis. For these graphs, called f-N curves, the value plotted on the y-axis is the discrete frequency of experiencing exactly N fatalities.
- Cumulative frequency basis. For these graphs, called F-N curves, the value plotted on the y-axis is the cumulative frequency of experiencing N or more fatalities.

Societal risk data and criteria are more commonly expressed in terms of cumulative frequency, thus F-N curves will be the principle focus of this appendix.

To construct the F-N curve, a list of all of the events (E_x) and their associated frequencies (f_x) and consequences (N_x) is compiled and sorted by decreasing value of N. In Table A.1, event E_1 is the most severe and has a predicted consequence of N_1, which occurs at a predicted frequency of f_1. The second most severe event is E_2, which has a predicted consequence of N_2, occurs at a predicted frequency of f_2, and so forth.

The equation in the last column of each row in Table A.1 calculates the cumulative frequency of occurrence for that event and all events preceding it in the table. For example, $F_3 = f_1 + f_2 + f_3$. Since the table was sorted by decreasing value of N (i.e., $N_1 > N_2 > N_3$), F_3 is the cumulative frequency of all events causing at least N_3 fatalities; in other words, the cumulative frequency of N_3 or more fatalities is $f_1 + f_2 + f_3$.

A.1.2   Calculating and Plotting an Example F-N Curve

To illustrate how societal risks may be presented as F-N curves, consider the results presented in Table A.2, representing a hypothetical QRA covering ten events that could result in multiple fatalities.

TABLE A.1. F-N Calculations

Event	Event frequency (per year)	Event consequence	Cumulative frequency (per year)
E_1	f_1	N_1	$F_1 = f_1$
E_2	f_2	N_2	$F_2 = f_1 + f_2$
E_3	f_3	N_3	$F_3 = f_1 + f_2 + f_3$
E_4	f_4	N_4	$F_4 = f_1 + f_2 + f_3 + f_4$
...	...	...	...
E_n	f_n	N_n	$F_n = f_1 + f_2 + f_3 + f_4 + \cdots + f_n$

TABLE A.2. Sample QRA Results

Event	Number of Fatalities	Frequency, f, per year
1	12	4.8×10^{-3}
2	123	6.2×10^{-6}
3	33	7.8×10^{-3}
4	33	9.1×10^{-4}
5	29	6.3×10^{-3}
6	16	7.0×10^{-4}
7	67	8.0×10^{-5}
8	10	4.0×10^{-3}
9	52	1.2×10^{-6}
10	3	3.4×10^{-4}

Rearranging the data in decreasing order of the number of fatalities gives the data set presented in Table A.3.

To put this data into the form needed for plotting an F-N curve, the frequency must be expressed in terms of the cumulative frequency. Table A.4 shows how this calculation is accomplished. Beginning at the top of the cumulative frequency (F) column (i.e., for the highest value of N), successively add the frequencies, moving down the column, so that any value of F is the sum of the f value to the left on that line and the F value on the line immediately above. The resulting data are presented in Table A.4 and plotted as an F-N curve in Figure A.1.

TABLE A.3. QRA Results Sorted By Decreasing N

Event	Number of Fatalities	Frequency, f, per year
2	123	6.2×10^{-6}
7	67	8.0×10^{-5}
9	52	1.2×10^{-6}
3	33	7.8×10^{-3}
4	33	9.1×10^{-4}
5	29	6.3×10^{-3}
6	16	7.0×10^{-4}
1	12	4.8×10^{-3}
8	10	4.0×10^{-3}
10	3	3.4×10^{-4}

TABLE A.4. F-N Presentation of QRA Results

Event	Number of Fatalities	Frequency, f, per year	Cumulative Frequency, F, per year
2	123	6.2×10^{-6}	6.2×10^{-6}
7	67	8.0×10^{-5}	8.6×10^{-5}
9	52	1.2×10^{-6}	8.7×10^{-5}
3	33	7.8×10^{-3}	7.9×10^{-3}
4	33	9.1×10^{-4}	8.8×10^{-3}
5	29	6.3×10^{-3}	1.5×10^{-2}
6	16	7.0×10^{-4}	1.6×10^{-2}
1	12	4.8×10^{-3}	2.1×10^{-2}
8	10	4.0×10^{-3}	2.5×10^{-2}
10	3	3.4×10^{-4}	2.5×10^{-2}

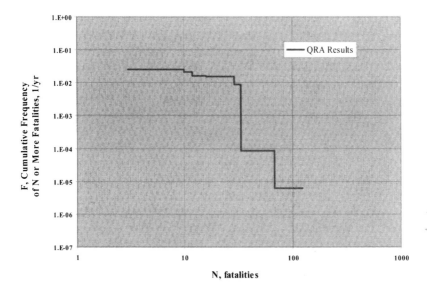

FIGURE A.1. F-N Curve for Hypothetical QRA Results

A.2   PRESENTATION OF SOCIETAL RISK CRITERIA

A.2.1   Defining the Risk Criteria F-N Curve

As discussed elsewhere in these *Guidelines*, societal risk criteria are generally presented as curves on F-N plots. Mathematically, the equation for an F-N criterion curve may be presented as: [Ball 1998]

$$F \times N^a = k \quad \text{or}$$

$$F = k \times N^{-a}$$

where,   F = the cumulative frequency of N or more fatalities
               N = the number of fatalities
               a = aversion factor (often between 1 and 2)
               k = constant

The slope of the societal risk criterion (when plotted on a log-log basis) is equal to -a and represents the degree of aversion to multi-fatality events embodied in the criterion. When the F-N curve slope is equal to -1, the risk criterion is termed 'risk neutral.' In this case, the risk criterion would dictate that the frequency of an event that results in 100 or more fatalities must be 10 times lower than the frequency of an event that results in 10 or more fatalities.

A risk criterion for which the curve slope is more negative than -1 is said to be more risk averse; in other words, the risk criterion reflects a greater concern for events causing larger numbers of fatalities. For example, if the F-N curve slope is equal to -2, the risk criterion would dictate that the frequency of an event that results in 100 or more fatalities should be 100 times lower than the frequency of an event that results in 10 or more fatalities.

An anchor point along the curve (e.g., N=10 fatalities, $F=10^{-3}$/year) and a slope (e.g., -1) is usually enough information to plot a risk criterion F-N curve. However, some organizations have developed societal risk criteria that use a moderate slope (e.g., -1) for low values of N, with an increase in slope (e.g., -2) for higher values of N. Such an example is included in Appendix C, Section C.2.

Figure A.2 shows the UK Health and Safety Executive's (HSE's) maximum tolerable societal risk criterion plotted with the hypothetical QRA results from Section A.1.2. As the figure illustrates, the QRA results would exceed the UK societal risk criterion for values of N less than about 30. In fact, if any portion of the calculated F-N curve exceeds the criterion line, the societal risk is said to exceed that risk criterion.

A.2.2   The Significance of the Slope of the F-N Curve

The underlying mathematics for F-N and f-N curves has been explored in a number of publications (see for example, [Schofield 1993] and [Hirst 2002]), and key points have been summarized in the report *Societal Risks*, by Ball. [Ball 1998] Ball points out that, while a slope that is more negative than -1 is often cited as being required to communicate risk aversion in an F-N curve, the allegedly 'risk neutral' F-N curve (slope = -1) actually exhibits a significant degree of aversion to multi-fatality events. This is illustrated in Figure A.3, which presents the f-N curve that can be back-calculated from the associated F-N curve.

The values of the discrete frequencies (i.e., f) were calculated noting that:

$$f_{x+1} = F_{x+1} - F_x$$

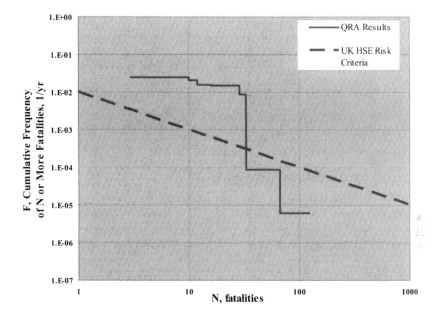

FIGURE A.2. Hypothetical QRA Results with UK HSE Societal Risk Criterion

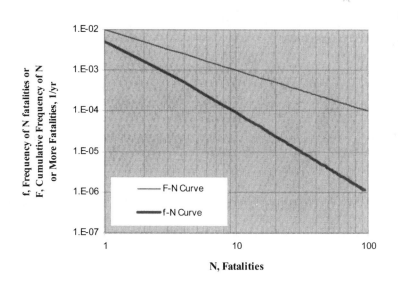

FIGURE A.3. Comparison of f-N and F-N Curves for A 'Risk Neutral' Slope

The F-N curve in Figure A.3 has a 'risk neutral' slope of -1, such as used in the UK HSE societal risk criterion curve. The slope of the f-N curve calculated from the F-N data is approximately -1.9 in the vicinity of N=10, increases to -1.99 at about N=100, and asymptotically approaches -2 as N increases. The discrete frequency associated with <u>exactly</u> 100 fatalities is nearly two orders of magnitude lower than the discrete frequency associated with <u>exactly</u> 10 fatalities. Thus, even the 'risk neutral' F-N curve has a significant degree of unrecognized aversion to multi-fatality events associated with it.

A.3   F-N CRITERIA SLOPE SELECTION

In a report sponsored by the HSE, Ball addresses in detail the issue of risk aversion and its role in F-N slope selection. [Ball 1998] Ball notes that many factors are involved in determining the public angst associated with major incidents. He also asserts that there are no conclusive indications that the scale of an incident, in terms of resultant fatalities, is the most significant determinant of public attitude about such events. Other factors, such as the public's unfamiliarity with the associated risk, come into play. For example, the public may assign more significance to another Three Mile Island-type nuclear reactor incident, with no fatalities, than they might to the next 300-fatality airplane crash.[20]

Ball also cites several studies of incident data indicating that the -1 slope more closely models the historic fatality data from the process industries. Similarly, Hirst and Carter, working with the Major Hazard Incident Data Service (MHIDAS) database, identified more than 1,600 incidents involving fatalities and created the F-N plot shown in Figure A.4. [Hirst 2002] The slope is approximately -1 over its entire length. Figure A.4 is not provided to suggest an historical basis for any particular F-N criterion curve slope. Indeed, it is not the intent, nor the purpose, of these *Guidelines* to recommend an F-N curve slope. The degree of risk aversion to be embodied in a societal risk criterion is, as discussed elsewhere in these *Guidelines*, a choice that an organization must make, ideally with regard to societal values. However, those considering reflecting a degree of risk aversion via the slope may want to consider the following:

- As shown in Section A.2.2, by virtue of the underlying mathematics, even an F-N slope of -1 has some inherent risk aversion.

---

[20] Refer to Slovic [Slovic 1987] for a more extensive treatment of public perception of risks.

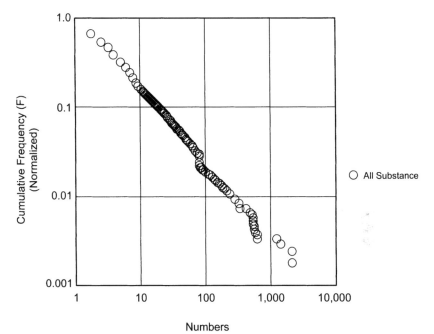

Numbers

FIGURE A.4. Historical F-N Curve with -1 Slope [Hirst 2002]

- Higher slopes potentially mean that more stringent risk controls (requiring greater resources) may be applied to reduce risk in the region where the confidence in the risk estimate is likely the lowest; in other words, where the cumulative risk is dominated by single remote events rather than combinations of smaller events. Unlike individual risk, societal risk decisions can be driven by some of the most remote likelihoods.

- Multiple fatality risk aversion says, in effect, that a fatality is more significant when it is accompanied by a large number of other fatalities. If preventing rare, high-consequence events requires disproportionate risk reduction efforts (in contrast with preventing more likely events), such efforts would run counter to a concept to which many risk analysts ascribe -- that the most effective allocation of risk reduction resources is that which maximizes the number of lives saved.

Appendix B
Survey of Worldwide Risk Criteria Applications

Chapter 3 detailed the evolution of risk criteria in the UK and in the Netherlands, and contrasted the practices in the two countries. The emphasis in the chapter was on discussing those precedents that a company might wish to consider when crafting its own criteria. This appendix expands the risk criteria survey to include other relevant criteria issued by governmental bodies at the country, regional, or local level and criteria issued by industry organizations.

The data in this appendix are a snapshot in time – regulatory precedents are in a state of continual flux. Organizations should seek to confirm and/or update any information presented here before using it as the basis for their risk criteria. Organizations working in a multi-national context may wish to refer to Section 4.13 regarding some of the unique challenges that they may face.

As noted previously, decisions related to the tolerability of risk can be made from two perspectives. The first relates to decisions focused on the risk-generating activity; for example, have adequate risk controls been implemented at the source? Such a question might be asked by a permitting body in the context of the review of a request to build, modify, or expand an industrial facility. It might be asked by a regulatory agency in the context of a requirement that the industrial facility file a safety study report, such as required by the Seveso Directive in Europe. Or it might be asked by the owners of the industrial facility seeking to responsibly manage the risk of their activities. This last context is the primary focus of these *Guidelines*.

The second perspective relates to decisions about the suitability of new or additional off-site developments in the vicinity of existing risk-generating industrial facilities; for example, is further development (such as new housing) prudent in the vicinity, given the current awareness of the level of risks posed by neighboring industrial facilities? Such a question might be asked in the context of land use planning.

The review in Chapter 3 focused on the first perspective, since the goal of that chapter was to establish the basis for the guidance provided in Chapter 4 on crafting company-specific risk criteria. However, since risk criteria precedents related to the second perspective, land use planning, also provide insight into societal values with respect to risk tolerance, criteria and practices related to land use planning are also discussed in this appendix – including those from the UK and the Netherlands.

B.1   EVOLUTION OF LAND USE PLANNING CRITERIA IN THE UK

B.1.1   Health and Safety Executive (HSE) - 1989

The HSE issued the document *Risk Criteria for Land-use Planning in the Vicinity of Major Industrial Hazards* in 1989. [HSE 1989b][21] The guidance was directed solely at decisions related to proposed new land use developments around existing major hazards. Issues related to the suitability of existing land uses or proposed new major hazards were explicitly excluded from the scope of the document. In making this distinction, HSE noted that valid reasons exist for taking a different approach to decisions about new major hazards as compared to new developments around existing major hazards. HSE observed that, generally, "… there may well be more scope for siting new developments a considerable distance from major hazards, than there is for siting new major hazards a considerable distance from any existing populations."

The planning guidance also explicitly excluded consideration of nuclear facilities, licensed explosives installations, and transportation activities (other than loading and unloading activities at covered major hazards sites).

In establishing criteria for land use planning decisions, HSE acknowledged that the process of making decisions about the significance of risk is "fraught with subjective perceptions." Consequently, HSE sought to incorporate societal values into the criteria:

> "… HSE has attempted to incorporate its understanding of the prevailing consensus based upon published views, decided cases, etc., and also leave a certain amount of flexibility."

---

[21] Note that all citations and quotations in Section B.1.1 are from [HSE 1989b].

In the implementation of land use planning in the UK, the HSE's role has traditionally been an advisory role to local planning authorities. After looking at the particulars of a situation referred to it, HSE would decide whether or not to recommend against the development. However, final authority for approving or denying the application rested with the planning authority.

HSE's approach to land use planning criteria is built upon its tolerability of risk (TOR) framework (see Section 2.7) and addresses both individual risk and societal risk considerations, but is less straightforward than the UK precedents outlined in Section 3.2.

Effect measures. In contrast to the UK risk criteria discussed in Section 3.2, the HSE decided against using fatality as the measure of harm upon which to base its land use planning criteria. In doing so, HSE noted:

"(a)   society is concerned about risks of serious injury or other damage as well as death;

(b)   there are technical difficulties in calculating the risks of death from a hazard to which individual members of a population may have widely differing vulnerabilities."

Instead, HSE introduced the concept of 'dangerous dose,' which it defined as an exposure to toxic gas, heat, or explosion overpressure that results in all of the following effects:

- severe distress would be experienced by almost everyone
- a substantial fraction of those exposed would require medical attention
- some people would be seriously injured, requiring prolonged treatment
- any highly susceptible people might be killed

Thus, for the HSE, land use planning decisions require an evaluation to determine the "...risk that a typical user of the development will be exposed to a dangerous dose or worse..." In identifying the typical user, HSE described two basic types of developments: 1) those used on a daily basis by the same people (e.g., homes, schools, and workplaces) and 2) those that might be used intermittently or only once by a particular individual (e.g., supermarkets, hotels). The first type of development would require consideration of both individual risk and societal risk, while the second type posed primarily societal risk issues.

Individual Risk Criteria. For individual risk criteria, HSE proposed that the lower bound (broadly acceptable) risk level should be a risk of 1 in a million (1×10^{-6}) per year of receiving a dangerous dose or more. For developments with a high proportion of highly susceptible people, HSE proposed a lower risk of 1/3 in a million (0.33×10^{-6}) per year of receiving a dangerous dose or more. For an upper bound (intolerable) risk level, HSE proposed a risk of 10 in a million (10×10^{-6}) per year of receiving a dangerous dose or more.

HSE estimated that a risk of 10^{-6}/year of receiving a dangerous dose or more would correspond to an individual risk of 10^{-6} fatality/year for the most vulnerable members of the population, and 0.33×10^{-6} fatality/year for the majority of the population. Similarly, a risk of 10^{-5}/year of receiving a dangerous dose or more would correspond to an individual risk of 10^{-5} fatality/year for the most vulnerable.

Societal Risk Criteria. HSE found the topic of societal risk to be more difficult to address. After reviewing the then-current state-of-the-art for consideration of societal risk, HSE decided against a quantitative (e.g., F-N based) approach to the consideration of societal risk in land use planning. Instead, the HSE elected to use qualitative judgments that combined individual risk information with the consideration of the number and vulnerability of the population in the proposed development. For example, the HSE might advise against a development of 30 houses if the individual risk was just above the 1 in a million per year risk of receiving a dangerous dose or more (i.e., the development met the quantitative individual risk criterion, but not the qualitative societal risk consideration).

Furthermore, HSE assigned greater significance to (and was more inclined to advise against) proposals that encroached on previously undeveloped land, as compared to incremental additions to existing developments. In contrast, HSE might not advise against a development where a small addition would be made to an existing population at risk.

HSE structured its qualitative societal risk guidance around the consideration of housing developments and then expanded the application to other types of developments by drawing equivalences, such as:

- 10 houses
- equal 100 people in a retail setting
- equal 100 people in a restaurant or pub
- equal 25 people in a hotel

Categories of developments. To implement its decision-making process, HSE defined four categories of developments, considering these factors:

- Inherent vulnerability of the exposed population (compare: health, age)
- Proportion of time spent by any individual in the development (compare: home, workplace, shopping, recreational facility)

- Population size
- Location indoors or outdoors, and ease of seeking shelter (compare: home, football stadium, theater, office complex)
- Ease of evacuation or other emergency measures
- Construction of buildings (height, ventilation, etc.)

Based on these criteria, the four categories of development were:

- Category A: Housing, hotel, or holiday accommodation – developments in use 24 hours a day where people live or are temporarily resident; population includes a mix of the healthy and unhealthy, young and old, and may be difficult to organize in an emergency.
- Category B: Some workplaces, parking areas, and so forth – includes factories, warehouses, offices, all with fewer than 100 occupants; and parking lots for fewer than 200 vehicles; occupants would tend to be fit and healthy, and could be organized easily for emergency action; general public will be present in small numbers and only for a short time in any day.
- Category C: Retail, community, leisure, and so forth – includes shops (250 m^2 or larger) and community facilities such as restaurants, pubs, or entertainment (100 m^2 or larger); members of the public will be present, but not resident; members of the public may be indoors or outdoors, and possibly in large numbers; emergency action may be difficult to coordinate. This category also includes cases not covered by A, B, or D; for example, housing with fewer than 10 units.
- Category D: Highly vulnerable or very large facilities – includes hospitals, homes for the elderly, schools, and so forth; also includes very large examples from category C (e.g., $5,000$ m^3 of floor-space, or provision for more than 1,000 people outdoors); institutional occupants may be especially vulnerable and difficult to move in an emergency.

Summary of criteria. HSE indicated that it would likely advise against developments in the following situations:

- "Housing developments providing for more than about 25 people where the calculated individual risk of receiving a defined 'dangerous dose' of toxic substance, heat, or blast overpressure exceeds 10 in a million per year.
- Housing developments providing for more than about 75 people where the calculated individual risk of receiving a 'dangerous dose' exceeds 1 in a million per year.

- For developments in category C ... HSE prefers not to suggest hard and fast rules based simply on risk figures. This is a category where the factors determining risk vary considerably in their importance, so it is difficult to generalize. HSE will tend to follow a similar approach to category A. Thus moderately sized developments will usually be inadvisable where the calculated individual risk for a hypothetical house resident exceeds 10 in a million per year, and large ones where the risk figure exceeds 1 in a million per year.

- The 1 in a million per year criterion may be extended downwards somewhat for developments which fall into category D, either because the inhabitants are unusually vulnerable or the development is extremely large.

- For all types of development, HSE is likely to advise more strongly against planning permission where the development forms a precedent for further growth or encroaches within any pre-existing cordon sanitaire, and comes within the risk criteria described above."

B.1.2   HSE – 2008[22]

Beginning in about 2003, the HSE undertook a fundamental review of its approach to land use planning, resulting in new guidance that was further updated in 2008 under the title *PADHI – HSE's land use planning methodology*. [HSE 2008d] The updated methodology is similar to, but more detailed than, that described above. In addition, the scope has been expanded to include consideration of hazardous material pipelines, explosives sites, and nuclear facilities.

Under the new guidance, the HSE establishes a consultation distance around major hazard sites and pipelines, based upon its modeling of the risks associated with the activity. The consultation distance is divided into an inner zone (IZ), middle zone (MZ), and outer zone (OZ). These zones and the consultation distance are illustrated in Figures B.1a and B.1b. The boundaries of these zones can be established based upon either a risk or consequence approach. For some facilities, the zones are established by iso-risk curves corresponding to the risks of receiving or exceeding a dangerous dose of 10×10^{-6}/year, 1×10^{-6}/year, and 0.3×10^{-6}/year, respectively. [HSE 2008e] For other facilities handling large amounts of flammable hydrocarbons, the zones

---

[22] The UK HSE released new land use planning guidance as these *Guidelines* were being finalized. [HSE 2007b] This new guidance, issued in response to the Buncefield fuel terminal explosion [MIB 2008], addresses concerns about the adequacy of prior methods of determining consultation distances in the vicinity of facilities storing large quantities of flammable liquids.

are established based upon threshold thermal dose or explosion overpressure values, without considering the likelihood of a release occurring. [HSE 2007b and ERM 2004]   A database of consultation distance information is maintained for use by the planning authority.

New criteria have been established for categorizing developments by sensitivity levels that, similar to the precedent set in [HSE 1989b], are determined based upon consideration of the number and vulnerability of the exposed population and other characteristics of the proposed development. Broadly, the four categories can be summarized as:

1 – People at work, parking
2 – Developments for use by the general public
3 – Developments for use by vulnerable people
4 – Very large and sensitive developments

HSE provides explicit definitions for a number of subcategories under each of these main classifications. [HSE 2008d]

Knowing the category of a particular proposed development, along with its location within any consultation distance surrounding an adjacent major hazard, allows the determination of HSE's recommendation based upon the decision matrix shown in Table B.1.   HSE's recommendation will be either Advise Against (AA) or Don't Advise Against (DAA).

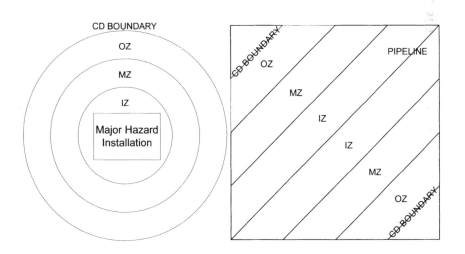

FIGURES B.1a. and B.1b. Illustration of HSE's Consultation Distance Around Fixed Facilities and Pipelines [HSE 2008d]

TABLE B.1. HSE's PAHDI Decision Matrix [HSE 2008d]

Level of Sensitivity	Development in Inner Zone	Development in Middle Zone	Development in Outer Zone
1	DAA	DAA	DAA
2	AA	DAA	DAA
3	AA	AA	DAA
4	AA	AA	AA

Detailed rules in the guidance address situations such as:

- The development straddles a zone boundary (i.e., is located in more than one zone)
- The development is exposed to more than one hazardous installation
- The proposal contains more than one development type
- An existing permitted use of the land is in place [23]

Under PADHI, the HSE maintains its advisory role, but automates the planning decision process via a website from which the planning authority can obtain information on consultation distances and, after inputting information on the proposed development, obtain the land use planning recommendation determined by the prescribed methodology.

B.2   LAND USE PLANNING CRITERIA IN THE NETHERLANDS

The situation in the Netherlands is more straightforward than in the UK, in that the same basic risk criteria are used for evaluating risk from both the perspective of the risk source and the risk receptor. In the Netherlands, under the *External Safety (Establishments) Decree* described in Section 3.3.7 [VROM 2004a], these considerations are addressed through the environmental permitting process (which reviews applications for new or modified industrial activities) and the spatial planning process (which reviews applications for new off-site developments). Local planning authorities, charged with the responsibility for implementing both processes, apply the current risk criteria in the conduct of both types of reviews.

The spatial planning process distinguishes between certain specified categories of establishments for which a standardized approach can be taken (e.g., liquefied petroleum gas [LPG] fueling stations, ammonia refrigeration installations, and facilities storing dangerous chemicals) and establishments where operations, and risk perspectives, are more complex (e.g., refineries and chemical processing plants). For the former establishments, safety distances

---

[23] The HSE has modified its guidance with regard to the existing use rule in the PADHI guidance (Rule 4c). Details are been provided in an interpretation letter from the Hazardous Installations Directorate. [HSE 2008f]

have been defined in the regulations, based on the 10^{-6} fatality/year individual risk criterion, and development may not be permitted within these zones. Furthermore, to ensure that the societal risk F-N criterion is not exceeded, maximum population densities have been established for the region between the 10^{-6} fatality/year individual risk contour and a maximum impact distance, defined by the 1% probability that an exposed individual would be fatally injured.

A quantitative risk assessment (QRA) is required for more complex establishments. A new risk source may not be permitted if this would cause the risk to existing residential populations to exceed the 10^{-6} fatality/year individual risk criteria for vulnerable populations, and new housing may not be permitted in an area if the risk from an existing industrial facility exceeds 10^{-6} fatality/year.

However, as discussed in Section 3.3.7, only the criterion for individual risk to vulnerable objects is binding; in other words, societal risk criteria and individual risk criteria for less vulnerable objects may be waived by the planning authority. A number of sources have pointed out that adherence to the established risk criteria has been problematic in the Netherlands (e.g., [VROM 2003], [VROM 2004b], [Ale 2005b]) with some noting that, of the two processes, compliance with the requirements of the environmental permitting requirements is the more rigorous.

B.3 HONG KONG

Hong Kong, concerned with high density residential developments located close to hazardous installations, mainly for bulk LPG storage, on Tsing Yi Island, commissioned a major hazards study in 1981. [Kolluru 1996] This and subsequent studies prompted the government to formulate preliminary risk criteria in the mid 1980s. [Ball 1998] The Coordinating Committee on Land-Use Planning and Control Relating to Potentially Hazardous Installations (CCPHI) was formed in 1986 to coordinate the government's risk management policy, and interim risk criteria were produced in 1988. [Ball 1998 and Boult 2000] The risk criteria were formalized and added to the *Hong Kong Planning Standards and Guidelines* [HKPD 2008] in 1993.

The individual risk and societal risk criteria are presented in Table B.2 and Figure B.2. The individual risk is a 'personal risk' and not a 'location risk'; in other words, the duration of exposure can be considered when calculating the individual risk. Hong Kong does not specify a broadly acceptable risk criterion for individual risk, nor does it specify the application of the As Low As Reasonably Practicable (ALARP) principle to individual risk. While some early risk studies proposed risk criteria for dangerous goods transport, no risk criteria currently exist for transportation risks.

QRA in Hong Kong is required by two different departments, the Planning Department and the Environmental Protection Department (EPD). The same risk criteria are used for both facility licensing and land use planning applications.

From a land use planning perspective, the Planning Department monitors compliance with the requirements for Potentially Hazardous Installations (PHIs), as documented in Chapter 12 of the *Hong Kong Planning Standards and Guidelines*. PHIs are defined as facilities that store hazardous materials in quantities equal to or in excess of established threshold quantities (TQs). These TQs were originally based upon the 1982 UK *Notification of Installations Handling Hazardous Substances Regulations*; however, some TQs have been revised to reflect considerations unique to Hong Kong. Examples of common PHIs and the associated substances and TQs are:

- LPG storage facilities, 25 tonnes
- Town gas installations, 15 tonnes
- Chlorine storage, 10 tonnes, or any storage in one tonne drums
- Gasoline (petrol) or naphtha storage, 10,000 tonnes
- Liquid oxygen storage, 500 tonnes
- Explosives factories/government storage depots, any quantity

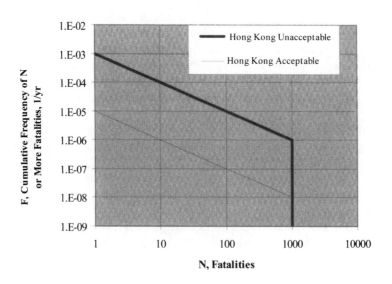

FIGURE B.2. Hong Kong Societal Risk Criteria

TABLE B.2. Hong Kong Individual Risk Criteria

Individual Risk Criteria, fatality/year	
Maximum tolerable risk to workers	N/A
Maximum tolerable risk to the public	
Existing situations	10^{-5} (See Note 1)
New situations	10^{-5}
Broadly acceptable risk	N/A

Note 1:   The risk criteria apply to both new and existing situations. However, more flexibility is apparent in the application of risk criteria to existing situations. See text for details.

Each identified PHI has a surrounding control zone, within which land use planning is required. Each PHI requires a QRA, which Hong Kong terms a hazards assessment. Based upon the results of the hazards assessment, a planning study is conducted to examine present and future land use and development proposals to identify appropriate planning considerations and development controls within the control zone. The risks to proposed developments are assessed against the risk criteria, as part of the application review.

From a facility licensing perspective, the EPD monitors compliance with the *Environmental Impact Assessment Ordinance* (EIAO). [HKEPD 2008a]   As outlined in a technical memorandum explaining the EIAO requirements, hazards assessments must be conducted for facilities that manufacture, store, use, or transport dangerous goods, if impacts from facility operation "may cause adverse public health effects … or pose an unacceptable risk to life." [HKEPD 2008b]

Examples of facilities that require environmental permits, and are therefore subject to requiring an environmental impact assessment, include but are not limited to: [HKEPD 2008a]

- A chemical or biochemical plant with a storage capacity of more than 500 tonnes and in which substances are processed or produced
- An oil refinery
- A petrochemical plant with an annual production capacity of more than 70,000 tonnes
- A bulk chemical storage facility with a storage capacity of more than 80,000 tonnes

Ball reports that Hong Kong followed UK precedents in developing its risk criteria [Ball 1998], which is not surprising considering its past history as a British colony. The societal risk F-N curve for maximum tolerable risk (10, 10^{-4}, -1) draws its anchor point from the Advisory Committee on Major

Hazards (ACMH) report. [HSC 1976] The curve for acceptable risk is two orders of magnitude lower. Both curves defining the societal risk criteria terminate at an N value of 1,000.

Full documentation of the rationale underlying the individual risk and societal risk criteria was not available. However, considering the low threshold quantities illustrated above for classification as a PHI, these risk criteria may prove to be hard to achieve for larger scale facilities. No discussion was found in the literature of the relative ease or difficulty experienced in applying these risk criteria to new, larger-scale facilities requiring Environmental Impact Assessments (EIAs). However, the planning process in Hong Kong appears to be effective in diverting new high-risk industrial developments to remote locations where meeting the societal risk criteria may be more readily achievable. In fact, the *Planning Standards and Guidelines* provide for proactively identifying and preserving remote areas as potential sites for future PHIs. [HKPD 2008]

While the risk criteria do not explicitly provide a distinction between new and existing facilities, the *Planning Standards and Guidelines* contain an implication of some flexibility in the land use planning application of the risk criteria with regard to existing facilities: [HKPD 2008]

> "Where the risk guidelines cannot be met, for existing PHIs, CCPHI will consider the necessary risk mitigation to bring the risk level down."

Discussion of the application of the ALARP principle is limited to societal risk, with the stated intent to "ensure that all practicable and cost-effective measures which can reduce risks will be considered." However, no mention is made of a gross disproportion test in the ALARP review. A less rigorous ALARP process would tend to offset the additional conservatism in the societal risk criteria.

Hong Kong has also proposed the use of QRA to evaluate the adequacy of controls to mitigate landslide risks. The Civil Engineering and Development Department has proposed risk criteria specific to this evaluation. [HK-CEDD 2007]

B.4 STATES OF SAO PAULO, RIO DE JANEIRO, AND RIO GRANDE DO SUL, BRAZIL

Three states in Brazil have implemented risk criteria: São Paulo, Rio de Janeiro, and Rio Grande do Sul.[24] The individual risk and societal risk criteria

---

[24] A federal regulatory agency, the Brazilian Institute of Environment and Renewable Natural Resources (IBAMA), requires QRA for environmental operating licenses. Reports indicate that IBAMA generally applies the São Paulo risk criteria.

for the three states are provided in Table B.3 and Figure B.3. Omitted from Figure B.3, for clarity, are the F-N curves for broadly acceptable risk, which are two orders of magnitude lower than the curve for maximum tolerable risk for São Paulo and one order of magnitude lower for the Rio de Janeiro and Rio Grande do Sul curves.

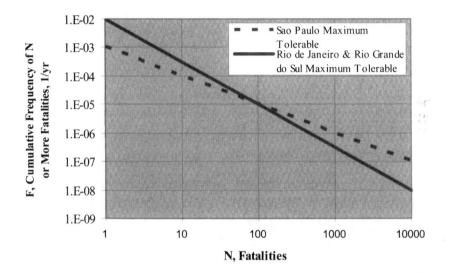

FIGURE B.3. Societal Risk Criteria for Three States in Brazil

As Table B.3 indicates, all three states also have individual risk criteria for pipeline transportation risks. São Paulo and Rio Grande do Sul have set these criteria one order of magnitude higher than the corresponding individual risks for facilities, while Rio de Janeiro has set its individual risk criterion for pipelines equal to the criterion for fixed facitities.

Rio de Janeiro is the only one of the three states that requires evaluation of societal risks for pipelines, again setting the criterion for pipelines equal to the criterion for fixed facilities.

TABLE B.3. Individual Risk Criteria for Three States in Brazil

Individual Risk Criteria, fatality/year	São Paulo, Rio Grande do Sul	Rio de Janeiro
Maximum tolerable risk to workers	N/A	N/A
Maximum tolerable risk to the public		
Existing situations	See Note 1	Plants: 10^{-5} Pipelines:10^{-5}
New situations	Plants: 10^{-5} Pipelines:10^{-4}	Plants: 10^{-6} Pipelines:10^{-6}
Broadly acceptable risk	Plants: 10^{-6} Pipelines:10^{-5}	N/A

Note 1: São Paulo: Risk criteria are applicable to new projects and significant modifications, and are applied to existing plants upon renewal of the operating license (the review frequency is generally 5 years, but varies with the type of activity being evaluated). Rio Grande do Sul: Risk criteria are not currently applicable to existing plants. Operating license renewal (every 4 years) is contingent on the update of the Risk Management Program of the company.

All three states require a preliminary consequence analysis to determine if a full QRA is required. State-specific triggers for the conduct of a QRA are described below.

São Paulo [CETESB 2003]
QRA guidelines, including risk tolerance criteria, have been issued by the state environmental agency, CETESB.[25] A full QRA is required if a loss of containment incident results in any of the following consequence endpoints reaching an off-site population:

- 0.1 bar overpressure for a vapor cloud explosion (VCE)
- Flammable vapor cloud with flash fire potential; in other words, assumed to be a concentration at or above one-half of the lower flammable limit (LFL)
- 1% fatality probability for toxics, based upon probit analysis, assuming a 10 minute exposure time

The QRA guidelines indicate the CETESB may, for fixed facilities, allow individual risk to exceed 10^{-5} fatality/year provided that the societal risk criterion is satisfied. QRA requirements, previously applicable to new projects and significant modifications to existing facilities, are now also being applied to existing facilities for routine renewal of operating licenses.

---

[25] Companhia de Tecnologia de Saneamento Ambiental

Rio de Janeiro [FEEMA, 2007]
The state environmental agency, FEEMA,[26] has not issued a standard QRA requirements document. However, requirements commonly imposed by FEEMA can be determined from terms of refernce issued to companies preparing to conduct invidual EIAs.

A QRA is required if the preliminary consequence analysis indicates that a flammable vapor cloud could reach a sensitive population, or if any such population is exposed to a 1% fatality probability from a fire, explosion, or toxic exposure (based upon a probit analysis). Sensitive populations would include residences, asylums, schools, prisons, penitentiaries, hospitals, and so forth.

Rio Grande do Sul [FEPAM, 2001]
QRA guidelines, including risk tolerance criteria, have been issued by the state environmental licensing agency, FEPAM.[27] A full QRA is required if a loss of containment incident results in any of the following consequence endpoints extending more than 500 meters from the release point:

- Immediately dangerous to life or health (IDLH) concentration for toxics
- 5 kW/m^2 radiant heat intensity for pool or jet fires
- LFL concentration for flash fires
- 13 kPa (0.13 bar) explosion overpressure (1% probability of eardrum rupture, for individuals outdoors)
- 7 kPa (0.07 bar) explosion overpressure (threshold for structural damage to residences, for individuals indoors)

The QRA guidelines indicate the FEPAM may, for fixed facilities, allow individual risk to exceed 10^{-5} fatality/year (to a maximum of 10^{-4} fatality/year) for surrounding industrial populations.

At this time, no information is available to provide a perspective on experience in the application of the risk criteria in these Brazilian states. Available information indicates that the application of risk criteria is focused on facility licensing issues, and not land use planning.

B.5   SINGAPORE

The Pollution Control Department (PCD) of the Singapore government has established QRA requirements for installations that store, transport, or use hazardous substances. [PCD 2007] Singapore's individual risk criteria are provided in Table B.4. Singapore has not established societal risk criteria.

---

[26] Fundação Estadual de Engenharia do Meio Ambiente
[27] Fundação Estadual de Proteção Ambiental

TABLE B.4. Singapore Individual Risk Criteria

Individual Risk Criteria, fatality/year	
Maximum tolerable risk to workers	N/A
Maximum tolerable risk to the public Existing situations	Industrial developments: 5×10^{-5} Commercial developments: 5×10^{-6} General public: 10^{-6}
New situations	Same as above
Broadly acceptable risk	N/A

As the table indicates, Singapore has implemented a three-tier perspective on individual risk for off-site populations by establishing maximum tolerable risk to: 1) industrial developments, 2) commercial developments, and 3) all other (general public). Singapore requires a 1 kilometer buffer zone between residences and installations such as oil refineries, petrochemical and chemical plants, and toxic industrial waste treatment facilities. [Singapore 2004]

Singapore's QRA guidelines also require that, for the worst-case credible scenarios, the following consequence criteria must be met: [28]

- 37.5 kW/m^2 thermal effect zone must not go outside the site
- 4 kW/m^2 thermal effect zone must not go outside industrial areas or encroach into residential areas or housing areas for construction workers
- Toxic effect zone corresponding to a 3% probability of fatality must not go outside industrial areas or encroach into residential areas or housing areas for construction workers
- 5 psi explosion overpressure zone must not go outside the site
- 0.5 psi explosion overpressure zone must not go outside industrial areas or encroach into residential areas or housing areas for construction workers

The QRA guidelines do not establish a risk criterion for broadly acceptable risk, nor do they discuss any application of the ALARP principle. Also, no guidance is provided as to whether the individual risk is to be calculated as a location risk or whether issues such as fractional time exposed are permitted to be reflected in the individual risk calculations. Reports indicate that the risk criteria are rigidly enforced by the Singapore authorities. [Lockwood 2007]

---

[28] Singapore defines a worst credible scenario as a scenario with the largest off-site hazard distance and an estimated frequency greater than 10^{-6}/year.

B.6   MALAYSIA

Malaysia requires that projects classified as major hazardous industrial installations submit a risk assessment with the EIA required by the Malaysian Environmental Quality Act.   Major hazardous industrial installations are defined as facilities storing or processing large quantities of toxic or flammable materials having the potential to harm surrounding population, property, or the environment. [Malaysia 2004]   The subject hazardous substances are listed in Schedules 1 and 2 of the *Occupational Safety and Health (Control of Industrial Major Accident Hazards) Regulations 1996.* [Malaysia 2006]
Malaysia has established the following individual risk criteria:

- For residential areas, schools, hospitals, and places of continuous occupancy: 10^{-6} fatality/year
- For industrial developments: 10^{-5} fatality/year

In addition, a buffer zone is required between the boundary of the major hazardous installation and any other development.   This buffer zone must be a minimum of 500 meters or the distance to the 10^{-6} fatality/year risk contour for the general public, whichever is greater.   Thus, the buffer zone effectively addresses land use planning considerations by ensuring that any development is beyond the 10^{-6} fatality/year risk contour.

B.7   AUSTRALIA

B.7.1   Western Australia

The Environmental Protection Authority of Western Australia published criteria for evaluating the risks posed by hazardous industries in 1987. [Western Australia 1987] These criteria, developed with public consultation, specified that an individual risk of 10^{-6} fatality/year for residential areas was "a risk level so small as to be acceptable to the Authority" and an individual risk of 10^{-5} fatality/year was "a risk level so high as to be unacceptable to the Authority."   Further evaluation was required for risks between these limits, constituting a de facto ALARP consideration.   In application, the Environmental Protection Authority generally required compliance with the lower 10^{-6} value fatality/year.
The individual risk calculation conservatively required the assumption that residents were outdoors, exposed to the risk 24 hours a day every day of the year, and would do nothing to avoid harm. In effect, the risk calculated was a location risk, as used in the Netherlands.
No societal risk criteria were proposed in the 1987 document.

In 1990, the Environmental Protection Authority sought to review and extend the risk guidelines. To enable public input, the Authority published a discussion document addressing issues such as individual risk, societal risk, and risk criteria for existing industries versus new industries. [Western Australia 1990]

Based upon considerable input from government agencies, companies, industrial groups, community groups, and individuals, the Authority modified its individual risk criteria in 1992 – notably, acknowledging other categories of off-site populations that were either more or less sensitive to risk exposure than residential populations. [Western Australia 1992a and 1992b] These criteria were further modified in 1994, in recognition that 'sensitive developments' such as hospitals might have 'risk generators' on site (e.g., LPG tanks) that should be considered in the risk profile. [Western Australia 1994] The resulting complement of individual risk criteria is reproduced verbatim below from a 2000 risk assessment guidance publication: [Western Australia 2000]

"(a)   A risk level in residential areas of one in a million per year or less is so small as to be acceptable to the EPA.

(b)   A risk level in "sensitive developments", such as hospitals, schools, child care facilities and aged care housing developments, of one half in a million per year or less is so small as to be acceptable to the EPA.

In the case of risk generators within the grounds of the "sensitive development" necessary for the amenity of the residents, the risk level can exceed the risk level of one half in a million per year up to a maximum of one in a million per year, for areas that are intermittently occupied, such as garden areas and car parks.

(c)   Risk levels from industrial facilities should not exceed a target of fifty in a million per year at the site boundary for each individual industry, and the cumulative risk level imposed upon an industry should not exceed a target of one hundred in a million per year.

(d)   A risk level for any non-industrial activity or active open spaces located in buffer areas between industrial facilities and residential areas of ten in a million per year or less is so small as to be acceptable to the EPA.

(e)   A risk level for commercial developments, including offices, retail centres, showrooms, restaurants and entertainment centres, located in buffer areas between industrial facilities and residential areas, of five in a million per year or less, is so small as to be acceptable to the EPA."

The list of sensitive developments given in criterion (b) was not intended to be comprehensive.

The calculated individual risk values are intended to be the summation of all risks from all facilities (exception: the 50×10^{-6} fatality/year industrial risk criteria), including the risk associated with any proposed new industry.

The above risk criteria are for new facilities. Where existing facilities pose risks exceeding these risk criteria, the intention is that risk reduction programs will be implemented with the ultimate goal of achieving the new criteria. The individual risk criteria are used both for evaluating proposed new facilities and for making land use judgments regarding new developments in the vicinity of existing facilities.

The Authority again elected not to propose societal risk criteria noting that "… experience with societal risk indicates that more research is needed before societal risk is addressed through the establishment of criteria." [Western Australia 2000]

Western Australia's individual risk criteria are summarized in Table B.5. As the table shows, risk criteria are proposed for a spectrum of populations, ordered below in terms of decreasing sensitivity:

- Sensitive developments, where large fractions of the population would be less able than the general population to protect themselves from the risk
- Residential areas
- Commercial developments
- Open areas and other non-industrial developments
- Industrial developments

TABLE B.5. Western Australia Individual Risk Criteria

Individual Risk Criteria, fatality/year	
Maximum tolerable risk to workers	N/A
Maximum tolerable risk to the public	
Existing situations	See Note 1
New situations	Sensitive developments: 0.5×10^{-6} Residential areas: 1×10^{-6} Commercial developments: 5×10^{-6} Other non-industrial activity: 10×10^{-6} Industrial activity: 50 to 100×10^{-6}
Broadly acceptable risk	N/A
Note 1:	Where the risks from existing facilities exceed the risk criteria for new facilities, risk reduction programs should be implemented with the ultimate goal of satisfying the criteria. See the text.

While Western Australia still intends that individual risk estimates be calculated as a location risk (i.e., assuming that someone is present 100% of the time), the graduated risk criteria appear to compensate for the varying ability of the different classes of population to protect themselves. Western Australia expects the implementation of a 'risk minimization' emphasis (i.e., application of the ALARP principle) in all cases, but does not require a gross disproportion test.

The individual risk for industrial developments should be interpreted to mean that no industrial facility should be exposed to a cumulative individual risk greater than 100×10^{-6} (sum of risks received from all other surrounding industrial facilities). Furthermore, the target facility should not see a risk of greater than 50×10^{-6} at its site boundary from any one of these other facilities.

In another guidance document, Western Australia proposes generic separation distances between industrial and sensitive land uses. For example, the specified separation around a chemical manufacturing facility could range from 300 to 1,500 meters. The intent is that these generic distances could be used to provide guidance in the absence of site-specific technical studies. However, Western Australia does not advise the use of these generic distances for major industrial developments. [Western Australia 2005]

Western Australia has not previously promulgated individual risk criteria for workers. However, implementation of the new *Dangerous Goods Safety (Major Hazard Facilities) Regulations 2007* [Western Australia 2007a] will require the development of risk criteria in support of the safety reports required by the regulation. Consequently, Western Australia has proposed both on-site individual and societal risk criteria. [Western Australia 2007b] These preliminary criteria are not discussed further here.

B.7.2   New South Wales

The New South Wales (NSW) Department of Planning (DOP) published criteria for evaluating the risks associated with major hazard facilities in 1990. [NSW 1990]   The individual risk criteria were, as a function of the exposed population:

- Hospitals, schools, child care facilities, old age housing: $\leq 0.5 \times 10^{-6}$ fatality/year
- Residential, hotels, motels, tourist resorts: $\leq 1 \times 10^{-6}$ fatality/year
- Commercial developments including retail, offices, and entertainment: $\leq 5 \times 10^{-6}$ fatality/year
- Sporting complexes and active open spaces: $\leq 10 \times 10^{-6}$ fatality/year
- Industrial: $\leq 50 \times 10^{-6}$ fatality/year

Comparing these to Table B.5 shows that these are essentially identical to the Western Australia risk criteria. The only distinction is the industrial risk criterion, which does not allow the increase to 100×10^{-6} fatality/year that Western Australia permits when an industrial site is exposed to externally generated risks from more than one source.

As was the case in Western Australia, NSW declined to craft societal risk criteria, noting that further research was required before NSW could address the issue. Instead, NSW proposed addressing societal risk in a qualitative fashion.

Citing a variety of considerations, including uncertainties in the analyses, NSW stated that the risk criteria should "be considered as providing target guidance rather than absolute values in all cases... [but that] substantial deviations from such targets should be fully justified." [NSW 1990] The risk criteria specifically applied to new installations and surrounding land use planning proposals, but the document included generalized guidance that, where existing situations exceeded the risk criteria, an individual risk of up to 10×10^{-6} fatality/year could be tolerable.

In 2007, DOP reaffirmed the individual risk criteria and provided additional guidance on their application, in a consultation draft of a new land use planning guidance document. [NSW 2007] In this document, DOP affirmed that:

- The risk criteria were to be applied to risk-based decisions involving both the development of hazardous installations and land use planning decisions addressing new developments around hazardous installations
- The individual risk was to be calculated as a location risk (assuming someone present continuously)
- Sensitive populations warranted risk criteria that were more conservative than residential populations, because of their inherently greater vulnerability
- The other classifications of population warranted less conservative risk criteria, reflecting that they were more able to take evasive action and/or were present a smaller percentage of the time
- The ALARP principle should be applied to 'avoid avoidable risk'

With respect to land use planning decisions, DOP proposed that no new or expanded developments should be permitted in the vicinity of hazardous facilities if this intensification of development would result in exceeding the risk criterion for the particular type of population.

The consultation draft acknowledged that existing hazardous facilities might face a greater challenge than a new facility in meeting the more restrictive risk criteria for residential and sensitive populations. The document appears to propose some relaxation of the risk criteria in certain situations; however, the details of the proposed guidance are not clear in the draft document and will not be reproduced here.

Societal risk criteria were included in the 2007 document, as shown in Figure B.4. The F-N curve for maximum tolerable risk draws its $(10, 10^{-4})$ anchor point from the ACMH report. [HSC 1976] However, it uses a slope of

-1.5, as did Rio de Janeiro and Rio Grande do Sul. The F-N curve for negligible risk is set two orders of magnitude below the F-N curve for maximum tolerable risk.

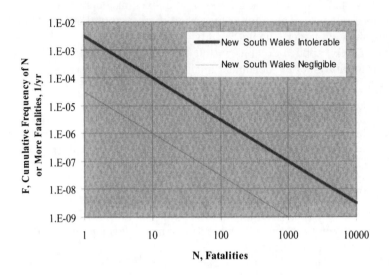

FIGURE B.4. New South Wales Proposed Societal Risk Criteria

Like Western Australia, NSW has a system of minimum separation distances between hazardous facilities and other types of developments that are intended to ensure that new developments satisfy the risk criteria. As noted above, and not unexpectedly, indications show that satisfying the risk criteria can be difficult where existing hazardous facilities and existing developments predate the land use planning controls – especially with respect to the more restrictive risk criteria for sensitive and residential populations.

Furthermore, the societal risk criteria "are indicative and provisional only and do not represent a firm requirement in NSW." [NSW 2007] That said, the document states that, if a significant portion of the societal risk plot lies in the ALARP region on the diagram, the "proposed development should only be approved <u>if benefits clearly outweigh the risks</u>" (emphasis added). Overall, the document appears to provide a good degree of flexibility with respect to implementation of the societal risk criteria.

Finally, NSW discusses the ALARP principle and cites the UK legal precedents associated with its application. [NSW 2007] However, the document does not clarify whether it is DOP's intent to interpret the ALARP principle as rigorously as does the HSE.

B.7.3 Queensland

In a 1998 information paper, *Hazardous Industry Planning for Safety*, the Chemical Hazards and Emergency Management Services (CHEM Services) branch of the Queensland government adopted the individual risk criteria, and underlying rationale, that had previously been adopted in NSW. [Queensland 1998] The NSW document cited was a 1992 reprinting of [NSW 1990]. The risk criteria are the same as those presented in Table B.5 and are illustrated in Figure B.5.

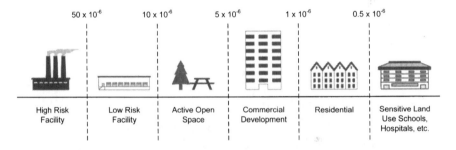

| 50×10^{-6} | 10×10^{-6} | 5×10^{-6} | 1×10^{-6} | 0.5×10^{-6} |

High Risk Facility | Low Risk Facility | Active Open Space | Commercial Development | Residential | Sensitive Land Use Schools, Hospitals, etc.

FIGURE B.5. Queensland Individual Risk Criteria [Queensland 1998]

CHEM Services reaffirmed the risk criteria in risk assessment guidance issued in 2002. [Queensland 2002] This document is important in that it explicitly communicates Queensland's intent that the ALARP principle be applied with the gross disproportion test.

Queensland, like Western Australia and NSW, has not mandated societal risk criteria, but suggests that "this concept should be addressed in the case of facilities that are close to significant population centres." [Queensland 1998]

No further discussion of the Queensland risk criteria will be provided, due to their overall similarity to the NSW risk criteria discussed previously.

B.7.4 Victoria

Risk criteria for both individual risk and societal risk were proposed by Victorian WorkCover in 1995, [Victoria 1995] and reaffirmed in a recent guidance note. [Victoria 2006] The individual risk criteria are repeated verbatim below:

- "Risk must not exceed 10 per million per year at the boundary of any new facility.
- If risk exceeds 10 per million per year at the boundary of an existing facility, risk reduction measures must be taken.
- If risk off-site is between 0.1 and 10 per million per year, all practicable risk reduction measures are to be taken, and residential developments are to be restricted.
- Risk levels below 0.1 per million per year are broadly tolerable."

The interim societal risk criteria are shown in Figure B.6. The F-N curve for maximum tolerable risk adopts the anchor point, $(10, 10^{-4})$ from the UK ACMH report, [HSC 1976] but uses the Dutch slope of -2. The curve for broadly acceptable risk is two orders of magnitude lower.

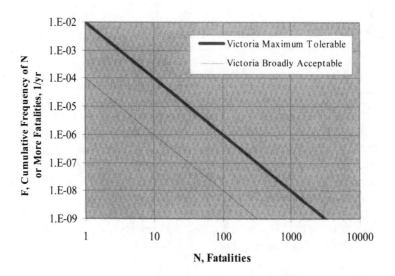

FIGURE B.6. Victoria Societal Risk Criteria

These interim criteria do not have any legislative standing and are advisory in nature. Facilities preparing safety cases under the Victoria major hazard facilities (MHF) regulation may use these risk criteria or develop their own.

Citing a 'continuous improvement' requirement in the MHF regulation, Victoria requires a very aggressive application of the ALARP principle. Victoria cautions against undue reliance on rigid criteria and requires that, even in the broadly acceptable risk region, "Risks must be managed and, so far as reasonably practicable, continuously reduced." [Victoria 2006]   In this sense, the approach to ALARP in Victoria is that, at least in theory, no risk level exists below which practical efforts at risk reduction should cease.

B.7.5   Other Australian States

No risk criteria currently exist for Tasmania, South Australia, Northern Territory, or the Australian Capital Territory.

B.8   CANADA

In 1994, the Major Industrial Accidents Council of Canada (MIACC) published risk criteria for land use planning decision making. [MIACC 1994] The criteria, summarized in Table B.6 and illustrated in Figure B.7, specified allowable land uses for defined levels of individual risk.   MIACC was an industry/government cooperative endeavor; thus, the MIACC risk criteria do not have any regulatory status and should be used only for guidance.   While MIACC has been disbanded, the risk criteria are described in a currently available document published by the Canadian Society for Chemical Engineering. [CSChE 2004]

TABLE B.6. MIACC Risk Criteria for Land Use Planning

Range of Risk Values, fatality/year	Permitted Land Use
>100 x10^{-6}	No land use other than the risk source itself (i.e., facility, pipeline). By implication, for on-site personnel, higher risks are acceptable.
10 x10^{-6} to 100 x10^{-6}	Uses involving continuous access and the presence of limited numbers of people, evacuation could be easily effected. For example, manufacturing facilities, warehouses, and open spaces.
1 x10^{-6} to 10 x10^{-6}	Uses involving continuous access and low population density, evacuation could be easily effected. For example, commercial uses, offices, and low-density residential areas.
< 1 x10^{-6}	Development is not restricted in any way, including institutional uses and high-density residential areas.

Annual Individual Risk

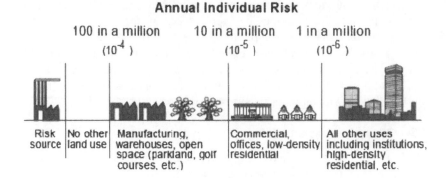

FIGURE B.7. MIACC Land Use Planning Risk Criteria [CSChE 2004]

The MIACC approach is similar to that described above for Queensland and Western Australia. While the risk criteria are intended to be compared to calculated individual risk values, the approach also addresses societal risk concerns, in that a graduated scale of population densities is built into the descriptions of the allowable land uses.

B.9   FRANCE

The French approach to major hazards risk has traditionally been deterministic; in other words, consequence based. However, in response to lessons learned from the Toulouse explosion, France enacted new legislation in July 2003 that addresses general principles of risk assessment, land use planning, risk communication, and compensation for the damages caused by accidents. This new law required, among other things, the conduct of risk assessments for industrial facilities and a renewed emphasis on land use planning. [France 2003] The land use planning requirements introduced a new tool – the technological risks prevention plan (in French: PPRT) that requires the identification of three risk zones around an industrial facility:

- Expropriation zone – in which the risk posed by the facility is so high that it warrants mandatory removal of residences
- Renunciation zone – a lower risk zone in which inhabitants have the right to request and receive compensation for voluntarily abandoning their residences
- Preemption zone – a zone in which the risk, while lower, is still significant and the community may wish to further reduce the risk by offering to buy out the inhabitants

A subsequent decree, issued in September 2005, defined the required structure and content for risk assessments, based upon a semi-quantitative approach. [France 2005] As the following will describe, decisions of whether to permit the construction of new or modified industrial facilities are based upon consideration of a societal risk measure, while land use planning decisions are based upon individual risk considerations.

To conduct the risk assessment, potential human health effects are first estimated and categorized according to effect thresholds for toxic, thermal (i.e., fire exposure) and overpressure (i.e., explosion) effects, corresponding to four intensity levels:

- Significant lethal effect (5% fatalities)
- First lethal effect (1% fatalities)
- Irreversible effect
- Reversible effect

A societal risk perspective is then obtained by considering the number of persons exposed when determining the consequence severity rating described by Table B.7. For example, a consequence severity rating of Extreme can result from more than 10 persons estimated to experience a significant lethal effect (5% chance of fatality) or from more than 100 persons estimated to experience a first lethal effect (1% chance of fatality).

TABLE B.7. French Matrix for Determining Severity of Consequences

Severity of Consequences	Significant lethal effect	First lethal effect	Irreversible effect
Extreme	PE > 10	PE > 100	PE > 1000
Catastrophic	$1 < PE \leq 10$	$10 < PE \leq 100$	$100 < PE \leq 1000$
Significant	$PE \leq 1$	$1 < PE \leq 10$	$10 < PE \leq 100$
Medium	0	$PE \leq 1$	$1 < PE \leq 10$
Moderate	No lethal effects outside the facility		$PE \leq 1$
Note: PE = Persons Exposed			

Frequency estimates for the assessed scenarios are expressed in ranges from A (most frequent) to E, as shown in Table B.8.

TABLE B.8. French Risk Assessment Frequency Ranges

Range Designation	A	B	C	D	E
Frequency, year^{-1}	$>10^{-2}$	10^{-3} to 10^{-2}	10^{-4} to 10^{-3}	10^{-5} to 10^{-4}	$<10^{-5}$

Based upon the estimates of frequency and severity of consequences, the planning authority can make a determination of whether authorization should be given to build a new facility or modify an existing facility. The risk matrix used to make this determination is shown in Table B.9.

TABLE B.9.   French Decision Matrix for Permitting New or Modified Facilities

Severity of Consequences	Frequency Range				
	E	D	C	B	A
Extreme	NO (new plant) RRM2 (existing plant)	NO	NO	NO	NO
Catastrophic	RRM1	RRM2	NO	NO	NO
Significant	RRM1	RRM1	RRM2	NO	NO
Medium			RRM1	RRM2	NO
Moderate					RRM1

In Table B.9, a NO indicates an unacceptable risk, and any proposed new facility or modification would be denied. RRM2 and RRM1 indicate the need for risk reduction measures, with the requirements for RRM2 being the more stringent.

The results of this exercise are also used to develop hazard level maps for use in land use planning decisions in the PPRT exercise. These maps are prepared by aggregating the risk information from all industrial facilities in the area, reflecting the fact that a particular area on the map may be impacted by multiple facilities. For each health effect intensity level and for each location on the map, the frequencies of the scenarios are summed. For example, if five scenarios have a frequency of E for a given intensity in a given place, the resulting probability will be 5E. Ten scenarios of the same frequency level are equivalent to one scenario in the immediately higher frequency level. A global hazard level is thus derived for each location on the map, as shown in Table B.10.

TABLE B.10. French Determination of Global Hazard Levels for PPRT

Maximum Effect Intensity Level	Significant lethal effect			First lethal effect			Irreversible effect			Reversible effect	
Sum of Frequencies	> D	5E to D	< 5E	> D	5E to D	< 5E	> D	5E to D	< 5E	>D	<D
Resulting Hazard Level	TF+	TF	F+	F+	F	M+	M+	M	Fai	Fai	Fai

In Table B.10, TF stands for very high (très fort in French), F stands for high (fort), M for medium, and Fai for weak (faible in French). Since the number of persons exposed did not factor into the global hazard level determination, the hazard level is reflective of individual risk. The use of the global hazard level in land use decision making is illustrated in Table B.11 for proposed new developments and in Table B.12 for existing land uses.

TABLE B.11. French Guidance for Proposed Land Use Decisions Based Upon Hazard Rating

	TF+	TF	F+	F	M+	M	Fai
Thermal and Toxic Effects	New Construction Prohibited				Authorized With Conditions	General Construction Permitted	
Overpressure Effects					Authorized With Conditions	General Construction	

TABLE B.12. French Guidance for Existing Land Use Decisions Based Upon Hazard Rating

	TF+	TF	F+	F
Expropriation	Mandatory for housing. Risk reduction options may be available for other land use activities.	Depending upon local debate.	Not Applicable.	
Renunciation	Not Applicable. See above.	Mandatory for housing. Risk reduction options may be available for other land use activities.		Depending upon local debate.

The 2003 legislation provided for the formation of local citizen advisory committees to work with planning authorities on certain aspects of the land use decision process. For example, in Table B.12 the committee might debate with the planning authorities to determine whether expropriation was the most appropriate option in the TF zone, or whether considering risk reduction measures would be more appropriate.

B.10 SWITZERLAND

The Swiss societal risk criteria for fixed facilities are depicted in Figure B.8. [SAEFL 1996] As the scale below the X-axis indicates, Switzerland has established criteria for a variety of consequences in addition to fatalities, implying a perspective on equivalency — an issue that some organizations have wrestled with and abandoned. Note that the fatality scale includes any deaths that occur within 30 days of an incident and as a direct consequence of the incident. The diagram reflects a cumulative frequency distribution, just as does a conventional F-N diagram.

The region of the F-N diagram to the left of a disaster value of 0.3 is termed the 'region of no serious damage' and includes fatality N values of up to 10. The three regions defined by the F-N curves, from top to bottom, are the unacceptable, transition, and acceptable risk regions. Gmunder describes the application of the criteria: [Gmunder 2002]

1. If the societal risk curve enters the unacceptable region, the facility must reduce the risk; otherwise, the enforcement authority can impose operational restrictions or require shutdown.
2. If the societal risk curve enters the transition region, the enforcement authority can define the tolerable level of residual risk, balancing the needs of the facility and the public, and the risk has to be reduced to that level.
3. If the societal risk curve is in the acceptable region for all values of N, the risk assessment procedure is complete. However, the facility is still expected to take all appropriate measures to further reduce risk – low levels of risk do not relieve the responsibility to seek further reasonable risk reductions.

Switzerland has also developed societal risk criteria for transportation risks. [SAEFL 2001] The criteria F-N curves and the multi-consequence scale for the N axis are identical to the fixed facility criteria; however, the unit on the frequency scale is fatality/year/100m.

Switzerland has not established an individual risk criterion.

B.11   SANTA BARBARA COUNTY, CALIFORNIA, USA

Santa Barbara County is one of the few local jurisdictions in the US requiring QRA and providing risk criteria for evaluating the results. QRAs are required by the county Planning and Development Department in matters involving the siting or modification of hazardous facilities or activities (e.g., hazardous material transportation), and other developments in the vicinity of a hazardous facility or activity.

The requirement for QRA is established in a safety element supplement to the County Comprehensive Plan. [Santa Barbara 2000] The risk criteria pertain only to acute public (off-site) risk, and are documented in the county's *Environmental Thresholds and Guidelines Manual*. [Santa Barbara 2006]

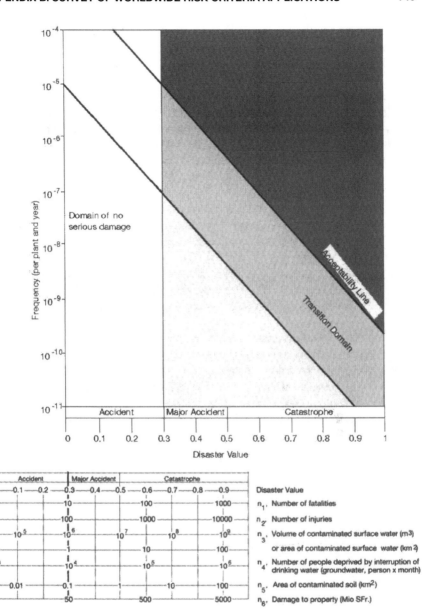

FIGURE B.8. Swiss Societal Risk Criteria [SAFEL 1996]

Each development plan requiring a land use permit must be evaluated to determine if a risk assessment is required. Certain types of hazardous facilities or activities are automatically subject to the QRA requirements (e.g., major sour gas pipelines). The need for a QRA for other types of hazardous facilities or activities is triggered (1) if a consequence assessment shows that a hazard zone extends to an off-site receptor and (2) if a preliminary risk screening yields an individual risk of 10^{-6} fatality/year or greater. [Santa Barbara 2006] The individual risk calculation can take into consideration the duration of exposure. Risk tolerability is judged based upon the societal risk, as discussed below. Individual risk is not a determinant of risk tolerability, only whether a detailed QRA is required.

The societal risk criteria are illustrated in Figure B.9. The risk criteria entail upper and lower bounding F-N curves, defined by (10, 10^{-5}, -2) and (10, 10^{-7}, -2). These curves establish three risk regions that, more typically, would be defined (starting from the top of the figure) as the intolerable, ALARP, and broadly acceptable risk regions. Santa Barbara defines these as the red, amber, and green zones, respectively. The following description of the application of Figure B.9 in risk decision making is taken from [Santa Barbara 2000]:

- New developments could be either a new hazardous installation or activity or a new non-hazardous development in the proximity of an existing hazardous installation or activity. For new developments, any risk profile (with a few exceptions) that, after mitigation, extends into the red zone is judged to be an unacceptably high risk warranting denial of the proposed development.

- A more stringent criterion is applied for highly sensitive land use, which is defined to include schools, hospitals, nursing home, stadiums, and so forth. Any new development involving highly sensitive land use with a risk profile that, after mitigation, extends into the amber zone is similarly judged to be an unacceptably high risk.

- Any risk profile that falls entirely in the green zone is deemed to have an insignificant impact on public safety and requires no mitigation.

- For existing facilities, any modification that increases risk, causing the risk profile to enter the red zone – or, for situations involving highly sensitive land use, to enter into the amber zone – would also be deemed an unacceptably high risk warranting denial of the proposed development.

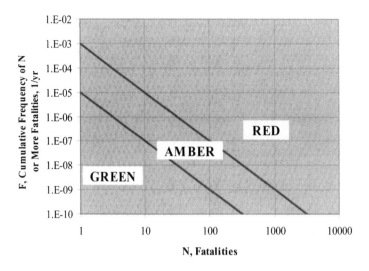

FIGURE B.9. Santa Barbara County Risk Criteria

The manner in which Figure B.9 is applied establishes, in effect, a risk criterion for highly sensitive land use that is two orders of magnitude lower than that for general land use. Santa Barbara also has risk criteria for injuries that are two orders of magnitude above the respective fatality F-N curves.

Apart from the stipulation that the risk being evaluated is to be 'mitigated risk', no inference of ALARP considerations exists in either of the Santa Barbara reference documents.

Santa Barbara notes that the risk criteria: [Santa Barbara 2000]

"… should not function as the sole determinants of significance for public safety impacts. Rather, they must be used in concert with applicable County policy, regulation, and guidelines to address other qualitative factors specific to the project which also help determine the significance of risk."

B.12   UNITED STATES FEDERAL GOVERNMENT AGENCIES

B.12.1 Nuclear Regulatory Commission (NRC)

The NRC's Probabilistic Risk Assessment (PRA)[29] Implementation Plan describes the various NRC activities to expand the use of PRA in the agency's regulatory functions. [NRC 1995] The NRC published a series of guidance

---

[29] The NRC uses the term probabilistic risk assessment (PRA) synonymously with the term QRA as it is used in these *Guidelines*.

documents describing how PRA should be used in changes to the licensing basis of nuclear power plants. Key principles and acceptance guidelines for this were published in July 1998 in Regulatory Guide 1.174. [NRC 1998]

NRC's approach to risk criteria is based on the following precept: [NRC 1986]

> "The risk to an average individual in the vicinity of a nuclear power plant of prompt fatalities that might result from reactor accidents should not exceed one-tenth of one percent (0.1 percent) of the sum of prompt fatality risks resulting from other accidents to which members of the U.S. population are generally exposed."

If the prompt fatality risk resulting from accidents to which the U.S. population is generally exposed is estimated as roughly 10^{-4} fatality/year (as shown in Table 4.2 in Section 4.9.1), the risk criterion would be 10^{-7} fatality/year. Note that this risk criterion addresses only the risk of public fatality. The use of the phrase 'prompt fatality' implies that the risk criterion does not reflect long-term fatalities that may occur many years after the accident (e.g., long-term health effects from the initial exposure to radiation or hazardous materials during an accident that ultimately result in death).

NRC's approach benchmarks the risk of catastrophic industrial (i.e., nuclear) accidents to other accident-related fatality risks commonly encountered and accepted by individual members of the public (e.g., motor vehicle operation or commercial air travel).

B.12.2 Department of Energy (DOE)

DOE's standards and guidance for the preparation of nuclear safety analysis reports [DOE 1991 and 1994] uses a risk matrix approach from which an underlying quantitative basis might be inferred. However, a more basic approach to a perspective on DOE risk criteria, with obvious linkage to the NRC precedent described above, is provided in a 1991 DOE directive: [DOE 1991]

> "The risk to an average individual in the vicinity of a DOE nuclear facility for prompt fatalities that might result from accidents should not exceed one-tenth of one percent (0.1%) of the sum of prompt fatalities resulting from other accidents to which members of the population are generally exposed. For evaluation purposes, individuals are assumed to be located within one mile of the site boundary.

The risk to the population in the area of a DOE nuclear facility for cancer fatalities that might result from operations should not exceed one-tenth of one percent (0.1%) of the sum of all cancer fatality risks resulting from all other causes. For evaluation purposes, individuals are assumed to be located within 10 miles of the site boundary."

Various DOE field offices have based explicit numerical risk criteria on the above precepts; for example: [DOE 1998 and 2004]

- A prompt fatality risk of 4×10^{-7} fatality/year (This is comparable to the NRC value above with the difference, presumably, due to difference in the database selected to establish background accidental fatality risks – DOE's value is actually more consistent with Table 4.2.)
- A latent cancer fatality risk of 2×10^{-6} fatality/year

DOE guidance for nonreactor facilities is addressed in DOE-STD-3009-94. [DOE 2006] The purpose of safety analysis for nonreactor facilities is not so much to quantify the risk, but to identify Design Basis Accidents with emphasis on identifying safeguards to be credited. DOE-STD-3009-94 states that events with an estimated frequency of occurrence of less than 10^{-6} fatality/year should be excluded from further consideration. These low frequency scenarios are considered in a qualitative process hazard analysis and, if judged to be 'Beyond Extremely Unlikely' (i.e., less than 10^{-6} fatality/year), are not carried forward to Accident Analysis (primarily a consequence analysis).[30]

At least one DOE field office has identified the need to fully evaluate the risk spectrum: [DOE 2002]

"To demonstrate conformance with these risk goals, the risk assessment should consider the risk of events that are less frequent than the extremely unlikely event category (i.e., with frequencies less than 10^{-6} per year)."

B.12.3 Environmental Protection Agency (EPA)

The EPA uses QRA to support decision making in the implementation of a number of regulations for which the Agency is responsible (e.g., to determine the degree of decontamination required for a waste site or the permissible levels of impurities in drinking water). In each case, the risk criterion pertains

---

[30] DOE's guidance on this point is in conflict with commentary elsewhere in these *Guidelines* that no scenario should be excluded *a priori* from a risk assessment because of a perceived low likelihood of occurrence.

to the risk to individuals of dying from cancer as a consequence of lifetime exposure to the chemical, termed by EPA as 'lifetime excess fatality risk'.

While risks associated with chronic chemical exposures are not within the scope of these *Guidelines*, the risk criteria associated with a number of regulations are summarized for contrast in Table B.13. [Rhomberg 1996] Note that the risk discussed is the risk of dying from cancer over a lifetime, which the EPA assumes to be 70 years (40 years for a working lifetime, for occupational exposures).   Refer to Section 1.2 for a discussion of the distinction between such chronic exposure risk criteria and the risk criteria for acute exposures presented in these *Guidelines*.   To contrast the EPA risk criteria with the risk criteria for acute exposures presented in these *Guidelines*, the values in Table B.13 must be divided by either 40 or 70.

TABLE B.13. Various Risk Criteria Used by EPA (as of 1996)

Regulation	Typical Acceptable Residual Risk, Fatality/lifetime	Comments
TSCA (Toxic Substances Control Act)	Occupational: 10^{-4} to 10^{-5} Non-occupational: 10^{-5} to 10^{-6}	Unstated, but values are typical
FIFRA (Federal Insecticide, Fungicide, and Rodenticide Act)	Occupational: 10^{-4} to 10^{-5} Non-occupational: 10^{-5} to 10^{-6}	Unstated, but values are typical
FFDCA (Federal Food, Drug, and Cosmetic Act)	Zero for additives Average diet: 10^{-6} Non-dietary: 10^{-6}	
SDWA (Safe Drinking Water Act)	10^{-4} to 10^{-6}	
CWA (Clean Water Act)	10^{-5} to 10^{-7}	
RCRA (Resource Conservation and Recovery Act)	For listing of a site: 10^{-5} Corrective action: 10^{-4} to 10^{-6} Incinerators: 10^{-5}	
CERCLA (Comprehensive Environmental Response, Compensation, and Liability Act)	10^{-4} to 10^{-6}	Depending in part on intended future use of site
CAA (Clean Air Act)	10^{-6}	Legislative trigger in the Clean Air Act Amendments for enhancement of Maximum Available Control Technology requirements

In some cases, the acceptable residual risk is presented as a range, such as 10^{-4} to 10^{-6} fatality/lifetime. The application of this range has been described as follows: [Commission 1997]

> "EPA ... has decided that it cannot use any single metric as a measure of whether a risk should be considered negligible. Instead, it has adopted a general presumption that a lifetime excess risk of cancer of about one in 10,000 (10^{-4}) for the most exposed person constitutes negligible risk and that the margin of safety should reduce the risk for the greatest possible number of persons to an individual lifetime excess risk no higher than one in 1 million (10^{-6})."

B.12.4 Food and Drug Administration (FDA)

The FDA, in addressing the safety of cosmetics and food additives, colors, and contaminants, uses in certain circumstances a risk criterion for lifetime cancer risk of 10^{-6}, presumable for a 70 year lifetime. [Rhomberg 1996]

B.12.5 Occupational Safety and Health Administration (OSHA)

In 1980, the Supreme Court invalidated an OSHA standard for workplace benzene exposure because OSHA had not demonstrated the risk associated with the 1 part per million (ppm) exposure limit that it had established. This, in effect, obligated OSHA to conduct QRAs to support the development of future exposure standards. With regard to how great a risk was required to be significant, Justice Stevens opined: [Rhomberg 1996]

> "If, for example, the odds are one in a billion..., the risk clearly could not be considered significant. On the other hand, if the odds are one in a thousand..., a reasonable person might well consider the risk significant and take appropriate steps to decrease or eliminate it."

While OSHA has not established an explicit criterion defining what is a significant level of risk, in application it appears that 10^{-3} fatality/year is the value commonly used. As with the EPA and FDA risk criteria, this is a lifetime cancer risk from chronic exposure. Again, refer to Section 1.2 for a discussion of the distinction between such chronic exposure risk criteria and the risk criteria for acute exposures presented in these *Guidelines*. Note, also, that OSHA assumes a 45 year 'career lifetime' for workplace exposure to chemicals, in contrast with the 40 year career lifetime assumed by EPA.

B.12.6 Department of Defense (DOD)

DOD has developed risk criteria for at least two applications.

Missile Test Ranges. Individual and societal risk criteria have been developed addressing hazardous activities associated with the operation of DOD missile test ranges. [DOD 2007a] As shown in Table B.14 and Figure B.10, the criteria address both the risks to the general public and to what the standard terms 'mission essential personnel' and 'critical operations personnel'. These groups are defined as:

- Mission essential personnel – Those persons necessary to safely and successfully complete a specific hazardous operation or launch.
- Critical operations personnel – This category includes persons not essential to the specific operation or launch currently being conducted, but who are required to perform safety, security, or other critical tasks at the range.
- General Public – People who are not declared/identified as mission essential personnel or critical operations personnel. This includes the public plus range personnel not essential to a mission, visitors, press, and personnel/dependents living on the base/facility.

Thus, the above definitions do not correspond precisely to the concepts of worker and public populations as used elsewhere in these *Guidelines*.

The values in Table B.14 require careful interpretation. The criteria expressed as probabilities (e.g., individual probability of fatality per event) represent an individual risk, but would have to be multiplied by the expected number of missions per year to obtain risk criteria such as used elsewhere in these *Guidelines*. The criteria expressed as expected casualties or expected fatalities represent societal risks. For example, for a given individual fatality probability, the expected number of fatalities will increase as the number of persons at risk increases. The expected fatality criteria in Table B.14 are analogous to the *average rate of death* societal risk measure described in Section 2.4.2.

TABLE B.14.   Individual and Societal Risk Criteria for DOD Test Range Operations

Basis	Undesired Event	Maximum Acceptable Value	
		General Public	Mission Essential and Critical Operations Personnel
Per Mission	Individual probability of casualty	1×10^{-6}	1×10^{-5}
	Expected casualties, casualty/event	1×10^{-4}	3×10^{-4}
	Individual probability of fatality	1×10^{-7}	1×10^{-6}
	Expected fatalities, fatality/event	3×10^{-5}	3×10^{-4}
Annual	Expected casualties, casualty/year	3×10^{-3}	3×10^{-2}
	Expected fatalities, fatality/year	1×10^{-3}	1×10^{-2}

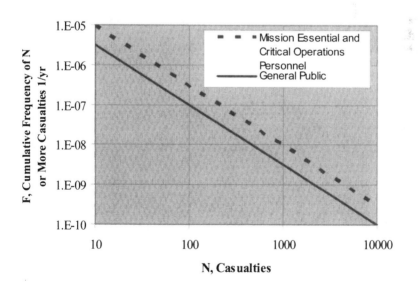

FIGURE B.10. DOD Test Range Societal Risk Criteria

The standard also provides more conventional societal risk criteria, as shown by the F-N curves in Figure B.10. DOD refers to these as catastrophic risk criteria, reflecting the potential for, and concern for, multi-casualty events. Note that the F-N curves in Figure B.10 use a slope of -1.5.

Also, note that DOD places emphasis on casualties, as opposed to fatalities. The societal risk F-N curve is expressed for casualties, and those individual risk criteria addressing fatalities are not required to be addressed in the range risk management program (i.e., they are identified as advisory requirements).

DOD Explosives Handling Operations. DOD risk criteria for explosives handling operations have been developed by the Risk-Based Explosives Safety Criteria Team (RBESCT) of the DOD Explosives Safety Board. [RBESCT 1999, available as an attachment to DOD 2007b] The criteria are summarized in Table B.15.

TABLE B.15. Individual and Societal Risk Criteria for DOD Explosives Handling Operations

Risk To:	Acceptance Criteria:
Workers	
Any one worker (annual probability of fatality), fatality/year	• Limit maximum risk to 1×10^{-4}
All workers (annual expected fatalities), fatality/year	• Attempt to lower risk if above 1×10^{-3} • Accept above 1×10^{-2} only with significant national need
Public	
Any one person (annual probability of fatality), fatality/year	• Limit maximum risk to 1×10^{-6}
All public (annual expected fatalities), fatality/year	• Attempt to lower risk if above 1×10^{-5} • Accept above 1×10^{-3} only with significant national need

As was the case with the range safety criteria, the criteria for any one worker or person represent individual risk criteria, while the criteria for all workers or all public represent societal risk criteria. The criteria for the public are intended to apply to the general public as well as government employees working at the facility but whose jobs are unrelated to the explosives activity.

B.13   NEW JERSEY, USA

The New Jersey Department of Environmental Protection (NJDEP) Air Quality Permitting Program (AQPP) requires a risk assessment when applying for an air pollution control pre-construction permit. The NJDEP individual risk criteria can be summarized as follows: [New Jersey 1994]

- If the incremental cancer risk from any contaminant evaluated is less than or equal to 10^{-6}, the risk is considered negligible.
- If the incremental cancer risk is greater than or equal to 10^{-4}, the risk is considered unacceptable.
- If the incremental cancer risk is between 10^{-6} and 10^{-4}, the risk is evaluated on a case-by-case basis.

The risk assessment addresses the incremental cancer risk to individuals associated with the lifetime exposure to chronic emissions, and parallels risk assessments described above for the US EPA. Thus, the risk criteria must be divided by 70 before comparisons can be, carefully, made to other risk criteria presented in these *Guidelines*.

B.14   INTERNATIONAL MARITIME ORGANIZATION (IMO)

In 1997, IMO published interim guidelines for the conduct of formal safety assessments (FSAs). [IMO 1997]   IMO defines an FSA as "a rational and systematic process for assessing the risks relating to maritime safety and the protection of the marine environment and for evaluating the costs and benefits of IMO's options for reducing these risks."   Missing from the interim guidelines, and the subsequently issued final guidelines [IMO 2002], were criteria for evaluating the risk.

Norway proposed individual risk and societal risk criteria for use in the FSA process in 2000. [IMO 2000]   While not yet formally adopted by IMO, these criteria were discussed extensively in a recent IMO publication on Risk Evaluation Criteria. [IMO 2005]   The individual risk and societal risk criteria are summarized in Tables B.16 and B.17, respectively.

TABLE B.16. Proposed IMO Individual Risk Criteria

Individual Risk Criteria, fatality/year	Existing Ships	New Ships
Maximum tolerable risk to workers (i.e., crew)	10^{-3}	10^{-4}
Maximum tolerable risk to the public (e.g., passengers or persons on shore)	10^{-4}	10^{-5}
Broadly acceptable risk	10^{-6}	10^{-6}

TABLE B.17. Proposed IMO Societal Risk Criteria

Tankers	Maximum tolerable risk	$(10, 2 \times 10^{-3}, -1)$
	Broadly acceptable risk	$(10, 2 \times 10^{-5}, -1)$
Bulk and Ore Carriers	Maximum tolerable risk	$(10, 10^{-3}, -1)$
	Broadly acceptable risk	$(10, 10^{-5}, -1)$
Passenger Roll On – Roll Off Ferries	Maximum tolerable risk	$(10, 10^{-2}, -1)$
	Broadly acceptable risk	$(10, 10^{-4}, -1)$

In defining the individual risk for existing ships, IMO chose the UK HSE criteria described in [HSE 2001]. Recognizing the potential for the implementation of newer technologies, IMO proposed setting the risk criteria for maximum tolerable risks for new ships one order of magnitude lower.

The societal risk criteria are all expressed as F-N curves having a -1 slope, with the anchor points given in Table B.17. As shown in Table B.17, the risk criteria are a function of the type of ship. To determine the anchor points for the various ship types, IMO proposed an approach premised on the assertion that "evaluation criteria may be associated with the economic importance of the activity in question, and calibrated against the average fatality rate per unit economic production." While the detailed mathematics will not be presented here, the approach can be outlined as follows: [IMO 2005]

1. From historical data, calculate an index defined as the ratio of the number of occupational fatalities to an associated measure of economic activity, such as the Gross National Product.
2. Determine the economic value of the activity being analyzed – in this case, the operation of a particular type of ship.
3. Calculate the tolerable average potential loss of life (PLL) for the activity (the product of the results of 1 and 2).
4. From the PLL, calculate the frequency, F, of one or more fatalities (N=1) for a F-N curve with a slope of -1 (the mathematics for doing so are described in Annex C of [IMO 2005]).
5. Set the maximum tolerable risk one order of magnitude higher and the broadly acceptable risk one order of magnitude lower.

While the concept described above may sound novel, Section 3.2.6 described a similar approach proposed by the UK ACDS for calculating local risk scrutiny levels for ports, based upon the risks at Canvey Island scaled by the relative scales of 'volume of trade'.

Comparison of the F-N curves described in Table B.17 shows them to be one to two orders less conservative than the UK public societal risk criteria F-N curves. Presumably, the ship's crew and passengers will be more aware of the hazards of the operation than would be the bulk of the off-site population around a chemical plant. And, the passengers demonstrate the concept of knowingly accepting a risk for the benefits derived from the risky activity.

The IMO approach to risk management provides for application of the ALARP principle. Analysis of historical risk data has shown that, for a variety of ship types, the historical risks are in the ALARP regions for the individual risk criteria and the appropriate societal risk criteria.   IMO cites this in support of its traditional emphasis on cost/benefit analysis in decisions related to potential risk reduction options. [IMO 2005]

B.15   OTHER NATIONS

The information for the nations discussed below was also noted in researching this Appendix. [EU 2005; EU 2006; Trbojevic 2004; and Trbojevic 2005] However, insufficient supporting information was available to permit fully developing these topics.

- The German approach to major hazards risk is to prohibit risk beyond the facility boundary.
- No national requirements exist for the conduct of QRA in Greece, Finland, Austria, or Spain.
- Hungary is reported to have off-site individual risk criteria of 10^{-5} fatality/year and 10^{-6} fatality/year for its upper and lower limits, respectively.
- The Czech Republic is reported to have risk criteria for maximum off-site individual risk of 10^{-5} fatality/year and 10^{-6} fatality/year for existing and new installations, respectively.
- Norway has required QRA for the off-shore industry since 1990, but requires the company to propose the risk criteria to be used. [Aven 2006]   The Norwegian Oil Industry Association has developed guidance on QRA, which includes some perspectives on the development of risk criteria. [NORSOK 2001]

B.16   SUMMARY

Tables B.18 and B.19 summarize relevant information from this Appendix and Chapter 3 for individual risk criteria for the public and for workers, respectively. Figure B.11 summarizes relevant public societal risk criteria.

TABLE B.18. Summary: Individual Risk to the Public [31]

Entities having two-limit system; i.e., Upper (intolerable) limit, Lower (broadly acceptable) limit and ALARA region between. Where different values for land use planning are established (e.g., separate criteria for industrial, commercial, and residential zones), the values for residential zones are used in the following table. Some risk criteria for certain land uses (e.g., "sensitive exposures") may be lower than those for residential.

	Entities/Applications	Comments
Upper Limit Values, fatality/year		
1×10^{-4}	State of São Paulo, Brazil/Pipelines	
	State of Rio Grande do Sul, Brazil/Pipelines	
	UK HSE/Fixed facilities and dangerous goods transport	
	International Maritime Organization (IMO)/Existing ships	Applies to both passengers and public ashore.
	State of Victoria, Australia	New facilities. If risk exceeds 10^{-5} fatality/year at the boundary of an existing facility, risk reduction measures must be taken. Non-mandatory, can be used as part of safety case.
	State of São Paulo, Brazil/Fixed installations	New installations and significant modifications to existing.
1×10^{-5}	State of Rio Grande do Sul, Brazil/Fixed installations	New installations.
	Hungary	
	International Maritime Organization (IMO)/New ships	Applies to both passengers and public ashore.

---

[31] Note: For completeness and comparison, some criteria for long-term health effects, such as cancer, have been included in these tables and have been clearly identified as such. The reader should carefully discriminate between these criteria and the criteria for acute health effects, which are the principle focus of these *Guidelines*.

Table B.18 Summary: Individual Risk to Public (cont'd)

	Entities/Applications	Comments
Upper Limit Values, fatality/year		
1.4×10^{-6}	New Jersey Air Quality Permitting Program	Value for unacceptable cancer risk to the public (10^{-4} cancer fatality/lifetime) divided by assumed lifetime of 70 years.
Lower Limit Values, fatality/year		
1×10^{-5}	State of São Paulo, Brazil/Pipelines	
	State of Rio Grande do Sul, Brazil/Pipelines	
	State of São Paulo, Brazil/Fixed installations	New installations and significant modifications to existing.
	State of Rio Grande do Sul, Brazil/Fixed installations	New installations.
1×10^{-6}	International Maritime Organization (IMO)/New and existing ships	Applies to both passengers and public ashore.
	UK HSE/Fixed facilities and dangerous goods transport	
	Hungary	
1×10^{-7}	Australia, State of Victoria/Fixed installations	Non-mandatory, can be used as part of safety case.
1.4×10^{-8}	New Jersey Air Quality Permitting Program	Value for negligible cancer risk to the public (10^{-6} cancer fatality/lifetime) divided by assumed lifetime of 70 years.

Table B.18 Summary: Individual Risk to Public (cont'd)

Entities having single-limit system:

Upper Limit Values, fatality/year	Entities/Applications	Comments
5×10^{-5}	Australia, state of Queensland	Specifies that 5×10^{-5} fatality/year risk contour must not extend beyond site boundary for new facilities. For existing facilities, risk reduction is to be "encouraged" if 5×10^{-5} fatality/year risk contour extends beyond site boundary.
	Singapore	Specifies that the 5×10^{-5} fatality/year risk contour may only extend into industrial development zones.
	Hong Kong	New installations. Existing installations exceeding this value should seek risk reductions.
	Netherlands	Applies to vulnerable objects. Existing situations. Interim value, existing situations must meet value for new situations (10^{-6} fatality/year, see below) by 2010.
1×10^{-5}	Canada, Major Industrial Accidents Council of Canada (MIACC)	Uses this value for low density residential, and lower value (10^{-6} fatality/year) for high density residential.
	Czech Republic	Limit for existing installations. Risk reduction must be carried out for facilities above this limit.
	State of Rio de Janeiro, Brazil/Fixed installations and pipelines	For existing facilities.
5×10^{-6}	Singapore	Specifies that the 5×10^{-6} fatality/year risk contour may only extend into industrial and commercial development zones.
2×10^{-6}	US DOE/Risk of latent cancer fatalities from nuclear facility accidents	Calculated from the criterion that risks should not exceed 0.1% of cancer fatality risks from all other causes.

Table B.18 Summary: Individual Risk to Public (cont'd)

Upper Limit Values, fatality/year	Entities/Applications	Comments
1×10^{-6}	Netherlands	Applies to (1) new permits for fixed installations, (2) new land use plans, and (3) transport of dangerous goods, including transport by pipelines.
	State of Western Australia, Australia	New installations. Higher limits are established for industrial and non-industrial developments. Lower limits (5×10^{-7} fatality/year) are established for "sensitive" exposures. Existing installations are to seek risk reductions to meet requirements for new installations.
	State of New South Wales, Australia	Higher limits are established for industrial and non-industrial developments. Lower limits (5×10^{-7} fatality/year) are established for "sensitive" exposures.
	State of Queensland, Australia	Higher limits are established for industrial and non-industrial developments. Lower limits (5×10^{-7} fatality/year) are established for "sensitive" exposures.
	State of Rio de Janeiro, Brazil/Fixed installations and pipelines	
	Czech Republic	Limit for new facilities.
	US DOD/Explosives handling activities	
	California, Santa Barbara County	Used as a screening value. Risk in excess of this value requires a risk assessment examining the societal risk from the facility.

Table B.18 Summary: Individual Risk to Public (cont'd)

Upper Limit Values, fatality/year	Entities/Applications	Comments
	Singapore	Specifies that the 1×10^{-6} fatality/year risk contour may only extend into industrial, commercial, and park developments zones.
	Malaysia	Malaysia: higher risk criteria established for those on industrial facilities, exposed from outside source.
4×10^{-7}	US DOE/Risk of "prompt" fatalities from nuclear facility accidents	Calculated from the criterion that risks should not exceed 0.1% of prompt fatality risks from all other accidental sources, assuming an accidental fatality rate of 4×10^{-4} fatality/year.
1.4×10^{-7}	Various US EPA regulations	Upper range value for cancer risk to the public (10^{-5} cancer fatality/lifetime) divided by assumed lifetime of 70 years.
1×10^{-7}	US NRC/Risk of "prompt" fatalities from nuclear power accidents	Calculated from the criterion that risks should not exceed 0.1% of prompt fatality risks from all other accidental sources, assuming an accidental fatality rate of 10^{-4} fatality/year.
1.4×10^{-8}	Various US EPA regulations	Lower range value for cancer risk to the public (10^{-6} cancer fatality/lifetime) divided by assumed lifetime of 70 years.
	Various US FDA regulations	Cancer risk to the public (10^{-6} cancer fatality/lifetime) divided by assumed lifetime of 70 years.

TABLE B.19. Summary: Individual Risk to Workers [32]

Entities having two-limit system: Upper (intolerable) limit, Lower (broadly acceptable) limit and ALARA region between.

Upper Limit Values, fatality/year	Entities/Applications	Comments
	UK HSE	
	IMO/Existing ships	For crew members on existing ships.
1×10^{-3}	State of Western Australia, Australia	Proposed for existing facilities. Where an existing facility exceeds 1×10^{-3} fatality/year, a risk reduction program with an agreed time frame must be implemented to achieve 10^{-3} fatality/year.
5×10^{-4}	State of Western Australia, Australia	Proposed for new facilities.
1×10^{-4}	IMO/New ships	For crew members on new ships.
Lower Limit Values, fatality/year		
1×10^{-4}	State of Western Australia, Australia	Proposed for new and existing facilities.
	UK HSE	
1×10^{-6}	IMO	For crew members on new or existing ships.

[32] Note: For completeness and comparison, some criteria for long-term health effects, such as cancer, have been included in these tables and have been clearly identified as such. The reader should carefully discriminate between these criteria and the criteria for acute health effects, which are the principle focus of these *Guidelines*.

Table B.19 Summary: Individual Risk to Workers (cont'd)

Entities having single-limit system:

Upper Limit Values, fatality/year	Entities/Applications	Comments
	US DOD/Explosives handling activities	
1×10^{-4}		
2.2×10^{-5}	Various US OSHA regulations	Cancer risk to workers (10^{-3} cancer fatality/lifetime) divided by assumed working lifetime of 45 years.
2.5×10^{-6}	Various US EPA regulations	Upper range value for cancer risk to workers (10^{-4} cancer fatality/lifetime) divided by assumed working lifetime of 40 years.
2.5×10^{-7}	Various US EPA regulations	Lower range value for cancer risk to workers (10^{-5} cancer fatality/lifetime) divided by assumed working lifetime of 40 years.

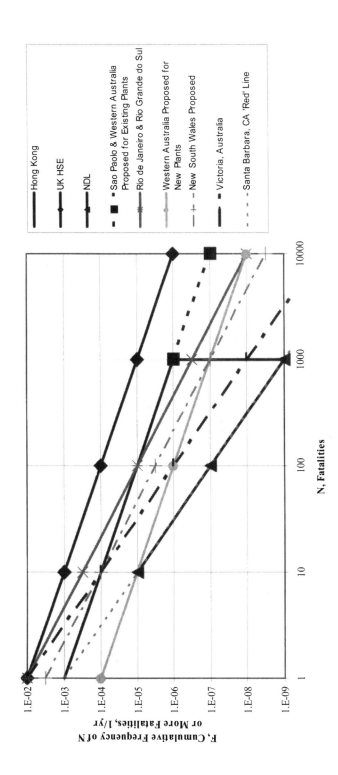

Legend:
- Hong Kong
- UK HSE
- NDL
- Sao Paolo & Western Australia Proposed for Existing Plants
- Rio de Janeiro & Rio Grande do Sul
- Western Australia Proposed for New Plants
- New South Wales Proposed
- Victoria, Australia
- Santa Barbara, CA 'Red' Line

N, Fatalities

F, Cumulative Frequency of N or More Fatalities, 1/yr

FIGURE B.11. Summary of Relevant Maximum Tolerable Societal Risk Criteria

Appendix C
Development of Company Risk Criteria

This appendix provides examples of how several companies have formulated their approaches to risk tolerance. The Center for Chemical Process Safety (CCPS) intends no implication of the suitability of these examples to any specific application.

C.1   CASE STUDY NUMBER 1

C.1.1   Development of Corporate Criteria

Company A is a major multi-national petrochemical company with operations in Europe, North America, and Asia.

In a review of the company's risk exposures, major incidents were identified as one of its key concerns. In order to better understand this issue, Company A decided to quantify the risks of major incidents potentially associated with all of its operations. Before carrying out this exercise, the company set out to establish risk criteria by which the significance of the calculated risks could be assessed.

Company A has a policy goal of zero incidents and consequently took the position that no level of risk would be deemed automatically acceptable. In line with its goal, all risks would be subjected to a process of continuing reduction, with risk mitigation prioritized based on the calculated level of risk.

In reviewing the experience of various other companies following major incidents, Company A recognized that large, multi-fatality events can have a significant impact on the long-term health, and even viability, of a company. This suggested that a certain level of societal risk existed above which the Company's long-term sustainability could be threatened; in other words, the company faced the possibility of losing its 'permission to operate' – either figuratively or, in some countries, literally.   Figure C.1 illustrates the company's conceptual approach to risk management and sustainability.

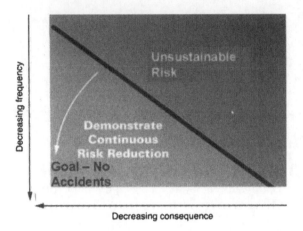

Decreasing consequence

FIGURE C.1. Company A's Perspective on Unsustainable Risk

Company A set out to establish this unsustainable level of risk by considering the impact on permission to operate at both a regional and a global level. It considered various sizes of incidents affecting the workforce and the general public, similar to those that had already occurred in industry. After careful consideration of such factors as voluntary versus involuntary risk, distribution of risks and benefits, and so forth, Company A concluded that differentiating between worker risk and public risk was appropriate, and it established risk criteria for each.

This section summarizes the overall approach for the company's development of its societal risk criteria.

Company A's operations are distributed across a number of separate business divisions. The business division leader has line responsibility for managing its performance (both financial and safety). Each business division is roughly equal in scale (in terms of financial turnover, number of employees, etc.). Company A therefore decided to set a level of unsustainable risk for each business division by dividing the Company unsustainable risk level by the number of business divisions, thus establishing the maximum tolerable risk for each business division. Since all business divisions will be required to operate below this maximum tolerable risk level, the total risk to the Company will necessarily lie below the corporate unsustainable risk level. This approach, coupled with the additional requirement for continuing risk reduction, was designed to provide long-term protection to the Company's permission to operate and to deliver its goal of zero incidents in the context to major incidents.

The societal risk criteria for a business division are expressed as two curves on a F-N graph – one for the workforce and another for the general public. The business division criteria are shown below in Figure C.2.

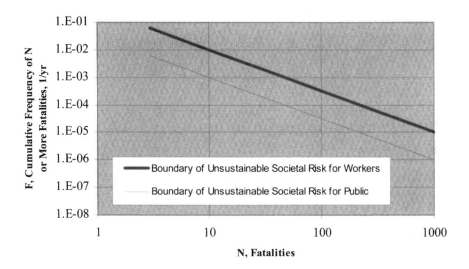

FIGURE C.2. Company A's Societal Risk Criteria for a Business Division

C.1.2   Applying the Criteria within the Businesses

The various business divisions differ widely in their operations.   Some business divisions contain a single major manufacturing site, whereas other business divisions consist of numerous much smaller sites, where blending, storage, and distribution activities predominate.

As an example of the former, business division #1's operations are focused on a single major site covering approximately 200 acres in land area with approximately 2,000 people on site. The site comprises:

- 12 manufacturing units
- 2 storage areas (raw material and finished products), with tank sizes ranging up to 10 million gallons capacity
- Road, rail, and ship loading/off-loading facilities
- A utilities area
- Numerous control rooms, workshops, warehouses, and an administration block

In performing its risk evaluation, business division #1 compared the risk of its single manufacturing site against the whole of the business division societal risk criteria.

In contrast, business division #2's operations are distributed across 100 specialty manufacturing/distribution sites. A typical site employs about 15 people and occupies an area of approximately 3 acres. Typical facilities would include:

- A small blending plant
- Road loading/off-loading facilities
- Drum storage
- A storage area with four tanks, each with a capacity of approximately 0.5 million gallons

To reflect the scale of each of its 100 individual sites, business division #2 divided the overall business division societal risk criterion by 100. The resultant societal risk criteria for sites within the two business divisions are shown in Figures C.3 and C.4.

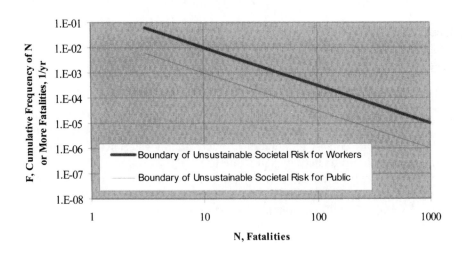

FIGURE C.3. Company A's Societal Risk Criteria for a Large Site

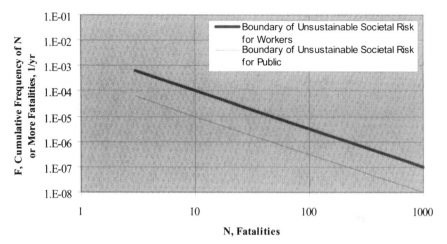

FIGURE C.4. Company A's Societal Risk Criteria for a Small Site

C.1.3   Validation of the Criteria against Regulatory Precedents

Since Company A operates in countries where risk criteria have been developed for use in a regulatory context, it decided to validate its approach against these regulatory precedents. Company A recognized that, if a more stringent national regulatory standard applied in a country where it operated, it would of course comply with that regulatory requirement.

Within the three major geographical areas where it operates, Company A has identified three relevant precedents for societal risk criteria for the general public (see Figure C.5). In researching further into the background of these criteria, Company A observed that:

- The UK criterion was based on consideration of the risks associated with a major oil and chemical complex (Canvey Island, consisting of ten existing and proposed refineries and major manufacturing/storage sites). The UK "intolerable" criterion for a single site corresponds to the societal risk criterion for the Canvey Island Complex divided by a factor of 10. This suggested that the UK criterion applied primarily to large sites, such as a major chemical manufacturing facility, a refinery, and so forth. Furthermore, Company A noted that the UK approach also relies strongly on the concept of As Low As Reasonably Practicable (ALARP), which typically drives risks well below the intolerable line.

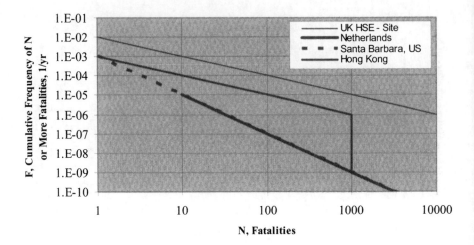

FIGURE C.5. Regulatory Criteria for Public Societal Risk Relevant to Company A's Business Portfolio

- The Hong Kong criterion has the same slope as the UK criterion, but is a factor 10 lower and was primarily intended to apply to small installations, such as liquefied petroleum gas (LPG) depots, water treatment works, and so forth.
- The Netherlands criterion applied to all installations (irrespective of size). Company A understood that, in practice, facilities that fall below the Netherlands criterion line were not subject to a strong push to drive risks lower.

These regulatory criteria were then compared against the company criteria as employed within the business divisions. Philosophically, the UK approach most closely aligns with that of Company A. In both cases, where risks fall below the "intolerable" level, a facility must still demonstrate that additional processes, such as ALARP or continuing improvement, have also been applied to drive risks lower. For the major manufacturing site in business division #1, Company A decided that the most appropriate comparison would be with the UK criterion. In contrast, for the smaller sites in business division #2, the Hong Kong criterion would be a more fitting comparison. Since the Netherlands criterion can apply to installations of all sizes, this comparison is also included for both business divisions. The comparisons are shown in Figures C.6 and C.7.

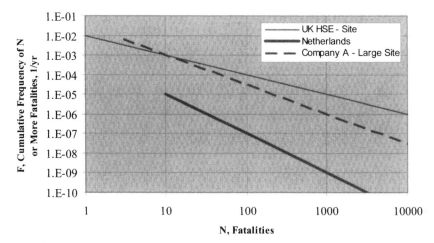

FIGURE C.6. Comparison of Company A's and Regulatory Criteria for Public Societal Risk for a Major Facility

Company A concluded that its societal risk criteria for the general public benchmarked satisfactorily in the regions in which it operates. In both cases, the criteria as applied in the business divisions were bracketed by the external regulatory precedents. As shown in Appendix B, no relevant regulatory precedents exist for societal risk for employees.

C.2   CASE STUDY NUMBER 2

C.2.1   Principles

Company B believes that regulatory agencies, governments, or industry associations working with the participation of the affected public and workforce can best determine what levels of risk are tolerable. Most of the relevant quantitative risk criteria have been issued by governmental bodies, often with significant public input and/or scrutiny, indicating a high degree of societal endorsement of the values. Therefore, as a principle, the risk criteria values used by Company B in its risk assessments are based upon worldwide regulatory and industry publications.

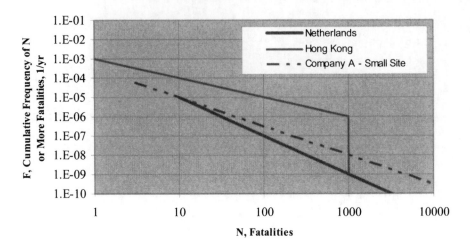

FIGURE C.7. Comparison of Company A's and Regulatory Criteria for Public
Societal Risk for a Small Facility

C.2.2   General Observations

Company B conducted a survey of the regulatory precedents before selecting
its risk criteria. This survey revealed three themes that, while not universally
applied, were factors in the development of many risk criteria.

1. A comprehensive risk management program should address both
   individual risk and societal risk.
2. Equitable risk criteria for the public should be lower (more
   conservative) than those for the workforce.
3. Even if a risk is below the criteria, management systems should be in
   place to drive further risk reduction as long as it is reasonable to do
   so.

Company B elected to address these three themes in its risk criteria.

C.2.3   Company Risk Criteria Based upon Regulatory Precedents

From its understanding of the basis of the UK Health and Safety Executive
(HSE) risk criteria, and considering that many Company facilities were of a

scale comparable to those envisioned by the UK HSE risk criteria, Company B elected to base its risk criteria on the UK precedents. Company B's risk criteria are shown in Table C.1 and Figures C.8 and C.9.

TABLE C.1. Company B's Individual Risk Criteria

Risk Criteria	Elevated Risk (As Determined by Published Regulatory Values and Practices)	Goal for Existing and New Facilities
Employees/Contractors (Fatality/Yr)	1×10^{-3}	$< 5 \times 10^{-4}$
Public (Fatality/Yr)	1×10^{-4}	$< 1 \times 10^{-5}$

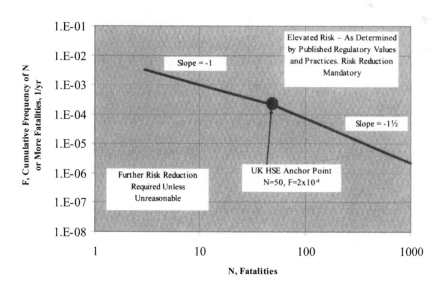

FIGURE C.8. Company B's Societal Risk Criteria for the Public

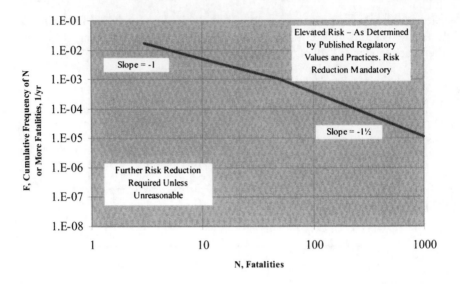

FIGURE C.9. Company B's Societal Risk Criteria for Employees and Contractors

C.2.4   Basis for Selection of Risk Criteria

Discussion of individual risk criteria. Tables B.18 and B.19 in Appendix B summarize the regulatory precedents for individual risk that were available for Company B's consideration as it was selecting its risk criteria. From the tables, Company B noted that UK HSE had proposed a maximum tolerable individual risk criterion for workers of 10^{-3} fatality/year, based upon the precedent that this value was the highest level of risk demonstrated for any industry that society was willing to accept. The UK individual risk criterion for the public was set one order of magnitude lower, at 10^{-4} fatality/year. Company B set more stringent goals of 5×10^{-4} fatality/year and 10^{-5} fatality/year for the workforce and the public, respectively, wanting to set higher standards for company operations. Risks below these criteria still require reasonable efforts for further risk reduction.

Discussion of societal risk criteria.   Figure B.11 graphs relevant regulatory precedents for societal risk criteria that were available for Company B's consideration as it was selecting its risk criteria. As with the individual risk criteria, Company B selected societal risk criteria based upon the UK HSE publications, extending the applicability to include employees and on-site contractor populations.

The UK HSE societal risk criterion curve for maximum tolerable risk to the public has its anchor point at 50 fatalities and a frequency of 2×10^{-4} fatality/year. For fewer than 50 fatalities, Company B's criterion utilizes an F-N slope of -1, in line with the values proposed by the UK HSE. For 50 or more fatalities, a greater negative slope (-1.5 versus -1) provides more stringent criterion than that of the UK HSE to help further reduce the potential for serious incidents.

Company B's societal risk criteria for employees and contractors is a conservative extrapolation of the UK HSE societal risk criteria for the public. Rather than an order of magnitude increase (ten-fold) in risk exposure for employees/contractors over the public exposures (as described above in the discussion of individual risk criteria), Company B established a 5-fold offset for employees and contractors.

C.3 CASE STUDY NUMBER 3

C.3.1 Development of Company Elevation Criteria

Company C is a multi-national chemical company with operations in many countries around the world. With a zero tolerance mindset for incidents, Company C developed a multi-faceted approach to risk assessment and management oversight. Inherent in this approach are protocols to promote appropriate communications – from the plant leader to corporate executives. Elevating data to the highest level of the company, as appropriate, is intended to ensure that appropriate steps are taken to ensure safety.

This Company has used quantitative risk assessment (QRA) as one of several internal risk assessment tools since the 1990s. Societal risk criteria were selected for evaluation of the QRA results based on a review of then-available criteria used by regulatory agencies and other industrial companies.

At the time Company C developed its criteria, no single or predominant criteria were in use within the process industries. Several of the then-available criteria used an upper criterion that included 10 fatalities at a frequency of 1×10^{-5}/year. Company C selected a linear criterion that passed through that anchor point, shown in Figure C.10 as Elevation Criterion #2. The slope of the risk criterion curve typically used by regulatory agencies and industrial companies ranged from -1 to -2; Company C chose a slope of -1.5, as shown in the figure. Company C observed that the criterion curve used by the Dutch government and several companies had a slope of -2. That slope yields a criterion that is more stringent at the high fatality end of the criterion curve and more lenient at the single fatality end. Several countries and companies also used a lower criterion curve (e.g., "broadly acceptable" criterion), typically two orders of magnitude below the upper criterion curve. Company C similarly established a parallel criterion curve that passes through the anchor point of $(10, 1\text{x}10^{-7})$, which is used as Elevation Criterion #1, to be described below.

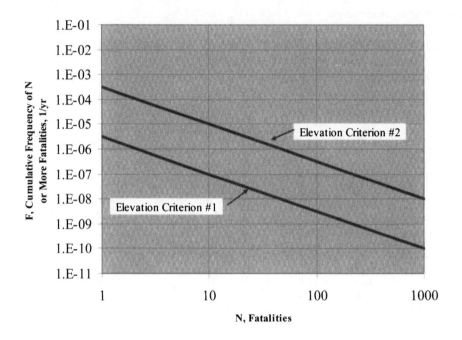

FIGURE C.10. Company C's Elevation Criteria

These criteria were used for evaluation of off-site risks projected by the QRA studies. Later, when implementing layer of protection analysis (LOPA), Company C adopted on-site criteria that were generally one order of magnitude higher than the off-site criteria. This is consistent with practices established by several countries and other companies.

C.3.2 Applying the Criteria

The criteria are applied across a wide range of chemical manufacturing sites of various sizes and technologies. The smallest site may contain only a single manufacturing unit and occupy just a few acres of property. The largest sites may contain multiple manufacturing units and occupy many acres of property. QRA studies and other risk assessments (e.g., LOPA) are typically performed on a single manufacturing unit (e.g., the unit to produce a particular substance). The risk criteria are used to elevate the level of management approval required to start up or continue to operate the site where the site's risk is projected to be above the criteria. Company C employs several levels of approval corresponding to the levels referenced in the above figure.

Typically, if the risks exceed the Elevation Criterion #1, a review of the risks is performed by the facility and global business unit leadership. If the risks exceed the Elevation Criterion #2, a review is conducted with corporate leaders or leadership teams. Upon review by the appropriate level of leadership, a decision is made about whether to continue operation of the evaluated facilities and, if so, under what conditions. Typically, the business leaders responsible for that facility are directed to implement additional measures to reduce risk. In several cases, this has resulted in the closure of plants, divestiture of the facility, or significant modifications.

C.3.3   Use of QRA and Risk Elevation Criteria

Company C uses a hierarchy of risk assessment tools, starting with qualitative or simple quantitative tools, then moving to semi-quantitative tools such as LOPA, and ultimately employing QRA if certain thresholds in the work process are exceeded. The level of risk review completed for each site is based upon the outcome of the previous level. If the responses from the first level reviews indicate a potential for relatively high risk, then the second level risk reviews (e.g., LOPA, explosion impact analysis) are completed. The LOPA studies consist of a semi-quantitative risk review of individual scenarios. If the LOPA evaluation shows that installing sufficient independent protection layers (IPLs) is impractical for any scenario, resulting in a higher than desired risk, a screening QRA study may then be performed. The screening QRA process uses sufficient information from all plant scenarios to estimate the individual risk created by that unit. If the individual risk values exceed a specified level, such as contours exceeding the plant fenceline, then a full QRA may be conducted to calculate the societal risk. When these QRAs are performed, the results are compared to the elevation criteria described above and appropriate actions are taken. This entire review process comprises a process risk management system and is described in a process risk management standard.

Since LOPA is the primary semi-quantitative risk assessment tool used for most process units, Company C's LOPA risk criteria for each individual scenario are aligned with the Elevation Criterion #1 described above. Although the elevation criteria were intended for use in evaluation of cumulative results across many scenarios, they are viewed as appropriate criteria for evaluating individual scenarios studied with LOPA. In a LOPA evaluation, the assumptions used for the frequency of initiating events and the probability credits allowed for mitigation features are intentionally conservative, given that the level of analysis typically does not consider factors considered in a QRA, such as wind direction. This is intended to ensure that the risk of the entire unit will be below the Elevation Criterion #2, even without the completion of a QRA, provided that each of the individual scenarios studied via LOPA are sufficiently mitigated.

Company C's criteria are subject to periodic review and change in the context of emerging practices in various countries and among various companies.

C.4   CASE STUDY NUMBER 4

C.4.1   Introduction

Company D is a large multi-national chemical company. Its worldwide operations range from low volume, high value specialty chemicals, often made in batch operations, to high volume petrochemical products made in large continuous plants. A wide range of hazardous materials is handled, including chemicals that have high acute inhalation toxicity, flammable liquids and gases, combustible powders, and reactive chemicals. The company has processes that operate at elevated temperature and pressure, and that involve highly exothermic reactions.

Company D was an early adopter of qualitative and quantitative process risk management techniques, including such tools as hazard and operability analysis (HAZOP), fault tree analysis, incident consequence modeling (fires, explosions, and toxic releases), and QRA. The initial implementation of these tools was primarily user driven. Tools were tried out in response to specific issues that arose on particular research or engineering projects, or in specific operating plants. The company realized that more effective tools were required to identify hazards, incident scenarios, and risk management strategies in technologies recognized as having high hazard potential. Such new tools were being developed in the industry and the company found them useful in understanding and managing process risk. The value of such tools was demonstrated by initial successes in identifying better strategies for process risk understanding and management. Accordingly, as usage increased, research and design teams and operating plant staff began requesting assistance in applying these tools.

C.4.2   Initial Use of Qualitative Tools

Qualitative process hazard analysis (PHA) techniques, primarily HAZOP, came to be commonly used on processes with significant hazards, including both batch and continuous processes – with the need for application of HAZOP to batch processes being an important requirement. As for most companies, no formal guidance was available on how to estimate the risk associated with the hazard scenarios identified, and the PHA team was left to determine if safeguards were adequate or if additional risk management features were required.

After some time, opportunities to improve the consistency of risk management decisions in qualitative studies became apparent. The company began to experiment with the use of qualitative risk matrices to evaluate risk and to determine where action was needed for risk management. After trying a variety of formats and risk categories, the company determined that employing a single, uniform risk matrix for general use in PHA studies would be most advantageous. The process safety group in the company reviewed internal experience with various matrices, reviewed available literature from external sources, and discussed the application of risk matrices with other companies. Company D then developed and implemented the first version of its internal risk matrix based on input from all of these sources. Use of the matrix was not required, but it was taught in PHA training and used in PHAs facilitated by the central process safety group.

With use, the matrix was modified and improved. In particular, it was expanded beyond the initial "safety" (injury and fatality) risk to consider both environmental and business (financial) damage. Eventually, after a number of years of use and improvement, the use of the risk matrix was incorporated into Company D's process safety management standard. By that time, it had become primarily a screening tool to determine how a particular risk should be managed – for lower risks, the initial PHA qualitative analysis was sufficient; for moderate risks, appropriate actions were required and the team would make a judgment if more detailed risk analysis was needed; and, for the highest risks, the use of a quantitative method (e.g., LOPA, fault tree analysis, QRA) was required to confirm that risk management expectations were being met.

C.4.3   Initial Use of Quantitative Tools

Company D was also an early adopter of QRA techniques. The company had several processes that involved highly exothermic reactions or complex continuous operations. These facilities were highly instrumented and had many process safety interlocks and alarms. Making a qualitative judgment of whether or not these safety systems were adequate was difficult. The company began to use fault tree analysis to understand the effectiveness of these systems for managing risk on several specific plants. Some significant successes were noted, with the analyses identifying hidden weaknesses in the complex system designs, as well as straightforward improvements that resulted in major risk reductions.

Initially, the studies were performed individually with the purpose of understanding how to improve a specific system. Risk decisions were made within the context of the particular study – the quantitative analysis identified opportunities for improvement and estimated the risk reduction that would result. Company engineers could then estimate the cost associated with the change and determine what should be done. In those early days of application of QRA tools to complex systems, significantly reducing risk at an affordable cost was nearly always possible.

As a result of this experience, Company D built up significant expertise in various QRA techniques, as well as a significant library of completed studies. Because QRA methods were specialized tools that required training and experience, and because fast, affordable computers were not generally available at that time, only a small group of people were involved in the initial studies. Consequently, a fairly consistent methodology was developed and applied.

Throughout this time, risk management decisions on specific projects were made in isolation (although many of the same people were involved in the decisions) and no effort was made to provide specific risk criteria. Company D was aware of various proposals to adapt quantitative risk criteria for use in evaluating the tolerability of residual risk remaining after the selected risk reduction measures had been applied. The company used this information informally in its internal process risk management decisions. This information included the technical literature on risk tolerance and early regulatory proposals for establishment of quantitative risk criteria. Most of the information used is found among the earliest references cited in these Guidelines.

After more than 10 years of experience with QRA techniques, Company D management decided to investigate the development of specific criteria to bring more consistency to risk decisions. QRA had proved itself to be a very useful risk decision-making tool, and its application was becoming practical for a wider range of projects and risk decisions. Computing power was increasing, and QRA tools were becoming more efficient and readily available. This meant that a broader community of people throughout the company would be able to use QRA tools. Since the practice of QRA and risk decision making would not be restricted to a small community of people expert in understanding and conducting QRA studies, additional guidance was needed to ensure uniform QRA practice and risk decision making.

C.4.4   Establishing Risk Tolerance Criteria

What resources were used in the process of establishing risk criteria? Company D had access to many internal and external resources that were valuable in establishing risk criteria. Some particularly valuable sources of information included:

- Previous QRA results and the risk management decisions made as a result of those studies. By the time it began to consider the establishment of risk criteria, Company D had been performing

QRA studies on selected facilities for more than 10 years. Company D had developed an internal risk analysis culture – a set of tools and methods that it used for QRA, a philosophy of how to deal with data uncertainty, and a set of data from past QRA studies that was used as a starting point for new studies. A database of QRA results was available, against which the proposed criteria could be tested. The hypothetical application of the new proposed criteria was expected to be generally consistent with decisions that had been made in the past by individual study teams. Because decisions based on prior studies had been made in the absence of criteria, a few exceptions to decisions that would have been made based on the criteria were expected. However, a large number of decisions inconsistent with the criteria would have suggested a need to reconsider the proposed criteria.

- Risk criteria published by other organizations and governments. These Guidelines provide many examples of publicly available information on risk criteria. At the time that Company D was developing its criteria, much less experience in this area and much less information was available. Also, summary documents collecting the available information were not available, and Company D had to collect the information itself through literature searches and discussions with other companies and government agencies. Company D found it extremely valuable to understand the numerical values of the risk criteria, the basis and thought process that the developing organization had used to specify the criteria, and the methodology that the developing organization expected to be used to develop risk estimates that would be compared to the criteria.

- Company D safety records and industry safety records. Proposed risk criteria should ensure that any facility that is tested against the criteria and determined to have tolerable risk will be at least as safe as the organization's past safety record. Furthermore, the risk criteria should provide a driving force for continuing improvement of that safety record. Company D used its own past performance data, and available industry data, to check on the reasonableness of its proposed criteria.

- Government and regulatory decisions when establishing regulations and safety and health standards. Because, at that time, little experience with acute risk criteria existed for process incidents such as fires, explosions, and toxic releases, Company D also considered less directly applicable government regulatory decisions as an indication of societal expectations – for example, decisions on the establishment of workplace exposure limits for certain toxic chemicals. Sometimes these decisions were based on some type of QRA, and they provided some guidance on society's expectations with regard to risk management.

- General societal risk data. These data provided some general context to Company D in establishing its criteria. However, one must consider the context of that societal risk data. Some particularly important factors include whether or not the risk is voluntary, and whether persons exposed to the risk derive any benefit from the activity that generates the risk. Societal risk data provide general background and context, but Company D regarded that such data should not be the formal basis for the establishment of risk criteria.

How was the information used to set risk criteria? Company D established its database of past QRA results and also a database of available risk decision-making criteria and societal risk data. The information on available risk criteria was studied, and the Company D team determined that it should set criteria for both individual and societal risk for off-site populations. The team selected proposed values based on an understanding of the available information from organizations and government agencies that had already established criteria, on the basis that these existing criteria represented the best available description of societal expectations. The values selected are provided in Table C.2.

TABLE C.2. Company D's Public Risk Criteria

Risk Measure	Units	Maximum Tolerable Risk	Negligible Risk
Individual risk at the location of a potential resident	Probability of fatality, per year	1×10^{-5}	$< 1 \times 10^{-7}$
Societal risk of 10 or more fatalities	Probability of an incident causing 10 or more potential fatalities, per year	1×10^{-5}	$< 1 \times 10^{-7}$

Company D then applied the proposed criteria to the results of past QRA studies to determine what kind of decision-making behavior would have been driven by the criteria. This also served as a check on whether the company's QRA processes and assumptions were appropriate to generate numerical results that were consistent with the proposed criteria. In general, Company D found that its QRA methodology was appropriate for use with the proposed criteria, and that the criteria, had they been in existence at the time of the past QRAs, would have resulted in decisions that were generally consistent with those made by the project teams and management based on their qualitative risk decisions.

Company D also compared its proposed risk criteria to its past safety record and found that the risks described by the proposed criteria were significantly lower than actual past experience. Therefore, the proposed risk criteria would drive improvement in safety performance; in other words, to meet the criteria, a facility would require an estimated risk significantly lower than the average past risk of a company facility. This makes sense considering how Company D allocates its resources – if a facility subject to a QRA meets the risk criteria, it is expected to have a significantly lower risk than the average company facility. Therefore, Company D should use its resources to identify and improve those company facilities that have above average risk, rather than continuing to try to drive risk even lower at a facility that is already perceived to be better than average.

Company D chose to apply the ALARP principle in developing its risk criteria. Its risk criteria had two parts – a numerical value above which it was necessary to have top management awareness and approval for continued operation, and a value two orders of magnitude lower, below which no additional risk mitigation action was required. Between the two values, plants or project teams were required to look for cost effective opportunities for risk mitigation, to evaluate them, and to implement them where they made sense. Company D's individual risk criteria are illustrated in Figure C.11.

Implementation. Company D implemented its risk tolerance criteria as a part of its corporate process safety management standard. This standard included guidance on risk methodologies, and specific requirements on how the risk tolerance criteria were to be used. In particular, a process was specified that established who was responsible for making and reviewing risk decisions made in accordance with the risk criteria. Although these criteria were developed many years ago, their existence has proven to be extremely valuable as risk analysis methodologies continue to evolve and become easier to use because of improvements in risk analysis technology and computing power. Risk criteria have been useful aids in implementing many new risk management techniques, far beyond what had ever been anticipated when the guidelines were initially developed – for example, risk-based inspection and maintenance and the design of safety instrumented systems.

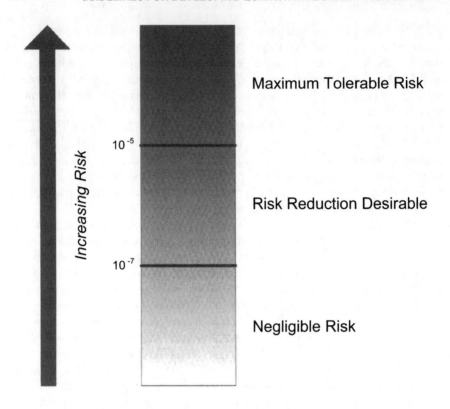

FIGURE C.11. Company D's Individual Risk Criteria, Illustrating Application
of ALARP

REFERENCES

ABS, 2000: *Guidance Notes on Risk Assessment Application for the Marine and Offshore Oil and Gas Industries,* American Bureau of Shipping, June 2000.

ADL, 1974: *A Modal Economic and Safety Analysis of the Transportation of Hazardous Substances in Bulk,* Arthur D. Little, Inc., U. S. Department of Commerce, Maritime Administration, Office of Domestic Shipping, May 1974.

Ale, 1984: *A computer based system for risk analysis of process plants,* Ale, B.J.M. and R. Whitehouse, Heavy Gas and Risk Assessment III, 5. Hartwig (Ed) D. Reidel, Dordrecht, The Netherlands, November 1984.

Ale, 1991: *Risk Analysis and Risk Policy in the Netherlands and the EEC,* J. Loss Prev. Process Ind., Volume 4, 1991.

Ale, 2005a: *Living With Risk: A Management Question,* Reliability Engineering and System Safety 90, pp. 196-205, 2005.

Ale, 2005b: *Tolerable Or Acceptable: A Comparison Of Risk Regulation In The United Kingdom And In The Netherlands,* Ale, B.J.M., Risk Analysis, Vol. 25, No. 2, 2005.

Arnauld, 1662: *Logic, Or the Art Of Thinking (La Logique, Ou l'Art de Penser),* Antoine Arnauld and Pierre Nicole, Port Royal Monastery, Paris, France, 1662. [A 1992 reprinting is available at: http://gallica2.bnf.fr/ark:/12148/bpt6k25788r. image.f1.langFR].

Aven, 2006: *Perspectives on Risk Acceptance Criteria and Management for Offshore Applications – Application to A Development Project*, Aven, Terje, et al., International Journal of Materials & Structural Reliability Vol.4, No.1, March 2006.

API, 2002*: Risk-Based Inspection,* API RP-580, American Petroleum Institute, Washington, DC, 2002.

API, 2003*: Management of Hazards Associated with Location of Process Plant Buildings,* API RP-752, American Petroleum Institute, Washington, DC, 2003.

Ball, 1998: *Societal Risks, Final Report,* D. J. Ball and P. J. Floyd, 1998.

Bekke, 2006: *Risk Criteria, Background information for maritime decision makers,* E.C.A. ter Bekke, Delft University of Technology, 2006. http://www.tudelft.nl/live/pagina.jsp?id=0f8a6698-1ebc-4a7b-b78c-043edc6c1645&lang=en&binary=/doc/risk_criteria.pdf

Bernstein, 1998: *Against The Gods – The Remarkable Story Of Risk,* Bernstein, Peter L., John Wiley & Sons, Inc., 1998.

BLS, 2008: Fatal Occupational injuries, employment, and rates of fatal occupational injuries by selected worker characteristics, occupations, and industries, 2006, US Bureau of Labor Statistics. http://www.bls.gov/iif/oshwc/cfoi/CFOI_Rates_2006.pdf

Bottelberghs, 2000: *Risk analysis and safety policy developments in the Netherlands,* Bottelberghs, P.H., Journal of Hazardous Materials 71, p. 59-84, 2000.

Boult, 2000: *Risk management of LPG transport activities in Hong Kong,* Journal of Hazardous Materials, 71, pp. 85-100, 2000.

Cambridge, 2008: Cambridge Dictionary of American English (on-line version) Cambridge University Press, 2008. http://dictionary.cambridge.org/define. asp?key=accept*1+0&dict=A

CCPS, 1992: *Guidelines for Hazard Evaluation Procedures (2nd Edition) with Worked Examples,* Center for Chemical Process Safety, New York, 1992.

CCPS, 1994: *Guidelines for Evaluating the Characteristics of Vapor Cloud Explosions; Flash Fires; and BLEVEs*, Center for Chemical Process Safety, New York, 1994.

CCPS, 1995: *Tools for Making Acute Risk Decisions with Chemical Process Applications*, Center for Chemical Process Safety, New York, 1995.

CCPS, 1996: *Guidelines for Evaluating Process Plant Buildings for External Explosion and Fires*, Center for Chemical Process Safety, New York, 1996.

CCPS, 2000a: *Guidelines for Chemical Process Quantitative Risk Analysis, Second Edition*, Center for Chemical Process Safety, New York, 2000.

CCPS, 2000b; *Evaluating Process Safety in the Chemical Industry – A User's Guide to Quantitative Risk Analysis*, Center for Chemical Process Safety, New York, 2000.

CCPS, 2001: *Layer of Protection Analysis – Simplified Process Risk Analysis*, Center for Chemical Process Safety, New York, 2001.

CCPS, 2007a: *Guidelines for Risk Based Process Safety*, Center for Chemical Process Safety, New York, 2007.

CCPS, 2007b: *Guidelines for Safe and Reliable Instrumented Protective Systems*, Center for Chemical Process Safety, New York, 2007.

CCPS, 2008: *Guidelines for Chemical Transportation Safety, Security, and Risk Management*, 2nd Edition, CCPS, 2008.

CEC, 2003: *Liquefied Natural Gas in California: History, Risks, and Siting*, California Energy Commission.
http://www.energy.ca.gov/reports/2003-07-17_700-03-005.PDF

CETESB, 2003: *Manual De Orientação Para A Elaboração De Estudos De Análise De Riscos*, P4.261, May 2003.
http://www.cetesb.sp.gov.br/Emergencia/ riscos/estudo/norma.asp

Commission, 1997: *Risk Assessment and Risk Management in Regulatory Decision-Making*, The Presidential/Congressional Commission on Risk Assessment and Risk Management, 1997.
http://www.riskworld.com/nreports/nr7me001.htm

CSChE, 2004: *Risk Assessment – Recommended Practices for Municipalities and Industry,* Canadian Society for Chemical Engineering, ISBN No. 0-920804-92-6, 2004.

Covey, 1990: The Seven Habits of Highly Effective People, Covey, Stephen R., Covey Leadership Center, 1990.

DNV, 2007: *QRA's for Dutch Installations. II. Overview of Risk Tools SAFETI-NL Selected by the Dutch Government,* Worthington, D. and Witlox, H., Loss Prevention and Safety Performance in the Process Industries, 2007, IChemE Symposium Series No. 153, IChemE, 2007.

DOD, 2000: *Standard Practice for System Safety,* MIL-STD-882D, U.S. Department of Defense, February 2000.
http://www.safetycenter.navy.mil/instructions/osh/milstd882d.pdf

DOD, 2007a: Common Risk Criteria Standards for National Test Ranges, Standard 321-07, U.S. Department of Defense, June 2007.
https://wsmrc2vger.wsmr.army.mil/rcc/manuals/321-07/321-07%20Common%20Risk%20Criteria%20Standards%20for%20National%20Test%20Ranges.pdf

DOD, 2007b: *Approved Methods And Algorithms For DOD Risk-Based Explosives Siting,* Technical Paper No. 14, Revision 3, Department of Defense Explosives Safety Board, U.S. Department of Defense, February 2007.
http://handle.dtic.mil/100.2/ADA463571

DOE, 1991: *Nuclear Safety Policy,* Secretary of Energy Notice SEN-35-91, September 9, 1991.
http://www.directives.doe.gov/pdfs/doe/doetext/senord/35/n3591.html

DOE, 1994: *Guidance for Preparation of DOE 5480.22 (TSR) and 5480.23 (SAR) Implementation Plans, DOE-STD-3011-94,* U.S. Department of Energy, November 1994.
http://www.osti.gov/bridge/servlets/purl/10194808-mY7ZJm/webviewable/10194808.pdf

DOE, 1998: *DOE Regulatory Unit Evaluation Report of the BNFL Inc. Safety Requirements Document,* RL/REG-98-01, U.S. Department of Energy, Richland Operations Office, 1998.

DOE, 2002: *Office Of Safety Regulation Position On Conformance With Risk Goals In DOE/RL-96-0006*, RL/REG-2000-08, U.S. Department of Energy, 2002.

DOE, 2004: *Top-Level Radiological, Nuclear, And Process Safety Standards And Principles For The RPP Waste Treatment Plant Contractor*, U.S. Department of Energy, DOE/RL-96-0006, 2004.

DOE, 2006: *DOE Standard, Preparation Guide for U.S. Department of Energy Nonreactor Nuclear Facility Documented Safety Analysis*, DOE-STD-3009-94, Change Notice 3, March 2006. http://www.hss.doe.gov/nuclearsafety/ns/techstds/standard/std3009/doe-std-3009-94_cn3_3-30-06.pdf

EU, 1997: *Council Directive 96/82/EC of 9 December 1996 on the control of major-accident hazards involving dangerous substances*, Official Journal of the European Communities No. L 10/13, 1997.

EU, 2005: SHAPE-RISK, Sharing Experience on Risk Management *(Health, Safety and Environment) to Design Future Industrial Systems*, Heidebrink, Ingrid, European Commission project No. NMP2-CT-2003-505555, February 2005.

EU, 2006: *Comparison Of Risk Analysis Methods And Development Of A Template For Risk Characterisation*, Ham, K, et al., European Commission – Joint Research Center, 2006.

Farmer, 1967: *Siting Criteria – A New Approach*, Farmer, FR, Atom, Vol. 128, pp 152-170. Also presented at the IAEA Symposium on Containment and Siting, Vienna, 1967.

Farquharson, 2003: *Using Quantitative Analysis to Make Risk-based Decisions*, 2003RM-047, Farquharson, McDuffee, RAMS Symposium, Tampa, FL, 2003.

Farquharson, 2007: *Facility Risk Review of a Uranium Processing Complex: A Risk-Informed Decision-Making Process*, Farquharson, J. A., et al., EFCOG 2007.

FEEMA, 2007: *Instrução Técnica Para Elaboração De Estudo De Impacto Ambiental – Eia E Seu Respectivo Relatório De Impacto Ambiental – Rima Para O Complexo Petroquímico Do Rio de Janeiro – Comperj*, Instrução Técnica Decon Nº 01/2007, Governo Do Estado Do Rio De Janeiro, 2007.

FEPAM, 2001: *Manual de Análise de Riscos Industriais,* FEPAM N.º01/01, Fundação Estadual de Proteção Ambiental, Brazil, 2001. www.fepam.rs.gov.br/central/formularios/arq/manual_risco.pdf

France, 2003: Loi n° 2003-699 du 30 juillet 2003 relative à la prévention des risques technologiques et naturels et à la réparation des dommages, Paris, 2003.

France, 2005: Arrêté du 29 septembre 2005 relatif à l'évaluation et à la prise en compte de la probabilité d'occurrence, de la cinétique, de l'intensité des effets et de la gravité des conséquences des accidents potentiels dans les études de dangers des installations classées soumises à autorisation, Paris, 2005.

Gmunder, 2002: *The Control Of Major Chemical Hazards In Switzerland In The Framework Of Sustainable Development – Liquefied Petroleum, Ammonia, and Chlorine As Examples*, Gmunder, F. K., et al., 6th International Conference on Probabilistic Safety Assessment and Management, June 23-28, 2002, San Juan, Puerto Rico. http://www.gmuender.org/stfv/ControlMajorHazards.pdf

Groningen, 1978: *Criteria for Risks Related to Dangerous Goods*, Provincial Policy Document on Environmental Issues for the Province of Groningen, the Netherlands, 1978.

Hendershot, 1995: *Risk guidelines as a risk management tool,* Dennis C. Hendershot, Process Safety Progress, Volume 15, Issue 4, pp. 213-218, 1995.

Hendershot, 1997: *No Good Deed Goes Unpunished: Case Studies of Incidents and Potential Incidents Caused by Protective Systems,* Process Safety Progress 16, 3 (Fall), pp. 132-139, 1997.

Hirst, 2002: *A "worst case" methodology for obtaining a rough but rapid indication of the societal risk from a major incident hazard installation,* Hirst, I.L. and Carter, D.A., Journal of Hazardous Materials, A92, pp. 223-237, 2002.

HK-CEDD, 2007: *Landslide Risk Management and the Role of Quantitative Risk Assessment Techniques,* Information Note 12/2007. http://www.cedd.gov.hk/eng/publications/information_notes/doc/in_200 7_12e.pdf

HKEPD, 2008a: *Environmental Impact Assessment Ordinance*, Hong Kong, 2008. http://www.epd.gov.hk/eia/english/legis/index1.html

HKEPD, 2008b: *Technical Memorandum to Environmental Impact Assessment Ordinance*, Section 4, Hong Kong, 2008. http://www.epd.gov.hk/eia/english/legis/memorandum/text4.html

HKPD, 2008: *Hong Kong Planning Standards and Guidelines, Chapter 12: Miscellaneous Planning Standards and Guidelines*, 2002. http://www.pland.gov.hk/tech_doc/hkpsg/english/ch12/ch12_text.htm

HSC, 1976: *The Control of Major Hazards, Advisory Committee on Major Hazards, First Report,* Her Majesty's Stationery Office, London, 1976.

HSC, 1984: *The Control of Major Hazards, Advisory Committee on Major Hazards, Third Report,* Her Majesty's Stationery Office, London, 1984.

HSC, 1991: *Major Hazard Aspects of the Transportation of Dangerous Substances,* Advisory Committee on Dangerous Substances, Her Majesty's Stationery Office, London, 1991.

HSE, 1978: *Canvey – An Investigation of Potential Hazards from Operations in the Canvey Island/Thurrock Area,* Her Majesty's Stationery Office, London, UK, 1978.

HSE, 1981: *Canvey – A Second Report,* Her Majesty's Stationery Office, London, UK, 1981.

HSE, 1988: *The Tolerability of Risk from Nuclear Power Stations,* Her Majesty's Stationery Office, London, 1988.

HSE, 1989a: *Quantified Risk Assessment: Its Input to Decision Making,* Her Majesty's Stationery Office, London, 1989.

HSE, 1989b: *Risk Criteria for Land-use Planning in the Vicinity of Major Industrial Hazards,* Her Majesty's Stationery Office, London, 1989.

HSE, 1992: *The Tolerability of Risk from Nuclear Power Stations (Revised),* Her Majesty's Stationery Office, London, 1992.

HSE, 2001: *Reducing Risks, Protecting People – HSE's Decision-making Process,* Her Majesty's Stationery Office, London, 2001.

HSE, 2004: *Development of an intermediate societal risk methodology, an investigation of FN curve representation*, Research report 283, HSE, 2004.

HSE, 2006a: *Safety Assessment Principles for Nuclear Facilities*, 2006. http://www.hse.gov.uk/nuclear/saps/saps2006.pdf

HSE, 2006b: *Offshore Installations (Safety Case) Regulations 2005 Regulation 12, Demonstrating Compliance With The Relevant Statutory Provisions, Offshore Information Sheet No. 2/2006*, 2006. http://www.hse.gov.uk/offshore/sheet22006.pdf

HSE, 2007a: *The Safety Report Assessment Manual, 2007.* http://www.hse.gov.uk/comah/sram/

HSE, 2007b: *Revised land use planning arrangements around large scale petroleum depots*, UK HSE, 2007. http://www.hse.gov.uk/research/rrpdf/rr511.pdf

HSE, 2008a: *HID's Approach To 'As Low As Reasonably Practicable' (ALARP) Decisions*, 2008. http://www.hse.gov.uk/comah/circular/perm09.htm

HSE, 2008b: *Guidance on 'as low as reasonably practicable' (ALARP) decisions in control of major accident hazards (COMAH)*, 2008. http://www.hse.gov.uk/comah/circular/perm12.htm

HSE, 2008c: *Principles and guidelines to assist HSE in its judgments that duty-holders have reduced risk as low as reasonably practicable*, 2008. http://www.hse.gov.uk/risk/theory/alarp1.htm

HSE, 2008d: *PADHI – HSE's land use planning methodology*, 2008. http://www.hse.gov.uk/landuseplanning/padhi.pdf

HSE, 2008e: *HSE'S current approach to land use planning (LUP)*, 2008., http://www.hse.gov.uk/landuseplanning/lupcurrent.pdf

HSE, 2008f: *PADHI Rule 4c (Existing Use Rule)*, Murray, John, HSE, 2008. http://www.hse.gov.uk/landuseplanning/rule4cletter.pdf

IMO, 1997: *Interim Guidelines For The Application Of Formal Safety Assessment (FSA) To The IMO Rule-Making Process*, MSC/Circ. 829 & MEPC/Circ. 335, International Maritime Organization, London, 1997.

IMO, 2000: *Formal safety assessment – decision parameters including risk acceptance criteria,* MSC 72/16, submitted by Norway, International Maritime Organization, 2000.

IMO, 2002: *Guidelines for Formal Safety Assessment (FSA) for use in the IMO rule-making process,* MSC/Circ. 1023, International Maritime Organization, London, 2002.

IMO, 2005: *Risk Evaluation Criteria,* D.4.5.2, International Maritime Organization, 2005.

Kletz, 2008: Personal communication with Trevor Kletz, 2008.

Kolluru, 1996: *Risk assessment and management handbook for environmental, health and safety professionals,* Kolluru, R., et al., New York: McGraw-Hill, 1996.

Lees, 2004: *Loss Prevention in the Process Industries (3rd Edition),* Mannan, S. (editor), Elsevier, 2004.

Lockwood, 2007: Personal communication, John R. Lockwood, 2007.

Malaysia, 2004: *Environmental Impact Assessment Guidelines for Risk Assessment,* Ministry of Natural Resources and Environment, Malaysia, 2004.

Malaysia, 2006: Occupational Safety and Health (Control of Industrial Major Accident Hazards) Regulations 1996, Malaysia, 2006. http://dosh.mohr.gov.my/koperat/LAW/Occupational%20Safety%20and %20Health%20Act%201994%20(Act%20514)/Regulations/pua039y19 96-cimah/pua0039y1996.pdf

MHIDAS, 2008: The Major Hazard Incidents Data Service.

MIACC, 1994: *Hazardous Substances Risk Assessment: A MiniGuide for Municipalities and Industry,* MIACC publication ISBN 1-89 5858-06-2, 1994.

MIIB, 2008: *The Buncefield Incident, 11 December 2005 - The final report of the Major Incident Investigation Board,* Buncefield Major Incident Investigation Board, 2008. http://www.buncefieldinvestigation.gov.uk/reports/index.htm

NASA, 1986: *Report Of The Presidential Commission On The Space Shuttle Challenger Accident, Appendix F,* National Aeronautics and Space Administration, Washington, D.C. June 6, 1986.

New Jersey, 1994: *Guidance on Preparing a Risk Assessment for Air Contaminant Emissions,* TECHNICAL MANUAL 1003, New Jersey Department of Environmental Protection, 1994.

NORSOK, 2001: *Risk and emergency preparedness analysis,* NORSOK Standard Z-013, Norwegian Technology Centre, Oslo, Norway, 2001.

NRC, 1975: Nuclear Regulatory Commission, WASH 1400, NUREG-75/014, *The Reactor Safety Study*, Rasmussen, 1975.

NRC, 1986: *10 CFR Part 50*, 51 Federal Register 30028 et seq., Nuclear Regulatory Commission, August 1986.

NRC, 1995: *Final Policy Statement on the Use of Probabilistic Risk Assessment Methods in Nuclear Regulatory Activities*, SECY-95-126, U.S. Nuclear Regulatory Commission, Washington, DC, May 18, 1995.

NRC, 1998: *Regulatory Guide 1.174, An Approach for Using Probabilistic Risk Assessment in Risk-Informed Decisions on Plant-Specific Changes to the Licensing Basis*, U.S. Nuclear Regulatory Commission, Washington, DC, 1998.

NSC, 2007a: *Report on Injuries in America, Selected Measures of Unintentional Injuries, U.S., 2001-2005*, National Safety Council, 2007. http://www.nsc.org/library/report_table_2.htm

NSC, 2007b: *What are the odds of dying?*, National Safety Council, 2007. http://www.nsc.org/lrs/statinfo/odds.htm

NSW, 1990: *Hazardous Industry Planning Advisory Paper No. 4 – Risk Criteria for Land Use Safety Planning,* New South Wales Department of Urban Affairs and Planning, 1990.

NSW, 2007: *Hazardous Industry Planning Advisory Paper No 10 - Land Use Safety Planning (Consultation Draft),* New South Wales Department of Planning, 2007.

PCD, 2007: *Guidelines for Quantitative Risk Assessment (QRA) Study*, Pollution Control Department, Singapore, 2007. http://www.nea.gov.sg/cms/pcd/qra_guidelines.pdf

Pfitzer, 2007: *A Comparison of QRA Methods used by DOD for Explosives and Range Safety with Methods used by NRC and EPA,* Pfitzer, B., et al., APT Research, Inc., 2007.
http://www.apt-research.com/pages/Papers/T-04-00200_ComparisonOfTheQRAMethodsWhitePaper.pdf

Queensland, 1998: *Hazardous industry planning for safety – a guide for industry and local government,* Chemical Hazards and Emergency Management Services, Queensland, 1998.

Queensland, 2002: *Guidelines for Major Hazard Facilities – C-Systematic Risk Assessment,* Chemical Hazards and Emergency Management Services, Queensland, 2002.

Rasmussen, 1975: *Reactor Safety Study, An Assessment of Accident Risks in U. S. Nuclear Power Plants,* Rasmussen, Norman et al., WASH-1400 (now NUREG-75/014), Washington DC: U.S. NRC, 1975.

RBESCT, 1999: *Criteria Selection for Risk-Based Explosives Safety Standards,* Pfitzer, Tom, et al., Risk-Based Explosives Safety Criteria Team, Department of Defense Explosives Safety Board, U.S. Department of Defense, 1999.

Rhomberg, 1996: *A Survey of Methods for Chemical Health Risk Assessment Among Federal Regulatory Agencies,* a report made to The Presidential/Congressional Commission on Risk Assessment and Risk Management, Lorenz R. Rhomberg, Harvard Center for Risk Analysis, 1996.
http://www.riskworld.com/nreports/1996/risk_rpt/pdf/rhomberg.pdf

Rijnmond, 1982: *Risk Analysis of Six Potentially Hazardous Industrial Objects in the Rijnmond Area – A Pilot Study,* Rijnmond Public Authority, D. Reidel, Dordrecht, the Netherlands and Boston, MA (ISBN 90-277-1393-6).

RSSG, 1993: *Risk Assessment. Report of a Royal Society Study Group,* The Royal Society, London, 1983.

SAEFL, 1996: *Evaluation Criteria for the Ordinance on Major Accidents, OMA, Evaluation of the Extent of Damage, Evaluation of the Acceptability of Risk, Guidelines for Establishments with Substances, Products or Special Wastes* (in French or German), Swiss Agency for the Environment, Forests and Landscape (SAEFL), Berne, Switzerland, 1996.

SAEFL, 2001: *Evaluation Criteria for the Ordinance on Major Accidents, OMA, Evaluation of the Extent of Damage, Evaluation of the Acceptability of Risk, Guidelines for Transportation Routes* (in French or German), Swiss Agency for the Environment, Forests and Landscape (SAEFL), Berne, Switzerland, 2001.

SAIC, 1995: *Shuttle Probabilistic Risk Assessment*, Washington, DC, Center for Aerospace Information, Science Applications International Corporation, 1995.

Santa Barbara, 2000: *Santa Barbara County Comprehensive Plan Safety Element Supplement*, Santa Barbara County, California, 2000.

Santa Barbara, 2006: *County of Santa Barbara Environmental Thresholds and Guidelines Manual*, Santa Barbara County, California, 2006. http://www.sbcountyplanning.org/PDF/ManualsReports/Manuals/Enviro nmental_Thrshlds.pdf

Schofield, 1993: *A Framework for Offshore Risk Criteria*, Schofield, S.L., Safety and Reliability, vol. 13, no. 2, 1993.

Singapore, 2004: *Code of Practice on Pollution Control (2000 Edition) (with amendments in Feb 2001, Jun 2002 and Feb 2004)*, Singapore, 2004.

Slovic, 1987: *Perceptions of Risk*, Science, Vol. 236, 1987.

TNO, 1983: *LPG Integral. Comparative Risk Analysis Of The Storage, Transshipment And The Use Of LPG And Benzene*, TNO-MT, Apeldoorn, 1983.

Trbojevic, 2004: *Risk Criteria in the UK and EU*, Vladimir M. Trbojevic, Presentation at the Workshop on ALARP and Societal Risk, Loughborough University, September 15, 2004.

Trbojevic, 2005: *Risk Criteria in EU*, Trbojevic, V.M., ESREL'05, Poland, June 27-30, 2005. http://www.risk-support.co.uk/B26P2-Trbojevic-final.pdf

Tweede Kamer, 1993: *Riscobenadering in het Milieubeleid; Verslag modeling overleg,* Tweede Kamer, 1993-94, 22666, nr. 5, 1993.

Victoria, 1995: *Interim Victorian Risk Criteria – Risk Assessment Guidelines,* DNV Technica, 1995.

Victoria, 2006: *The Requirements for "Demonstration" Under the Occupational Health and Safety (Major Hazard Facilities) Regulations,* Major Hazard Facilities Regulations Guidance Note, MHD GN-16, 2006.

Voogd, 2004: *Disaster Prevention in Urban Environments,* Voogd, Henk, European Journal of Spatial Development, no. 12, Sept. 2004.

VROM, 1984: *LPG Integral Memorandum (Circulaire LPG Integraal Nota),* Netherlands, Staatsuitgeverij, The Hague, 1984.

VROM, 1988: *Major Incident Risk Order (Besluit Risico's Zware Ongevallen, in Dutch),* Staatsblad, 1988, 432, The Hague, 1988.

VROM, 1989: *Premises for Risk Management (Omgaan met Risico's, in Dutch),* Tweede Kamer, vergaderjaar, 1988/1989, 21137, nr.5.

VROM, 1995: *A Review of Risk Control,* M.J.Pikaar, M.A.Seaman, nr. 1995/27A & 27B, The Netherlands, 1995.

VROM, 1999: *Hazards of Major Accidents Decree (Besluit Risico's Zware Ongevallen, in Dutch),* Staatsblad, 1999, The Hague, 1999.

VROM, 2003: *Summary of the advice on external safety published by the Advisory Council for Transport and Water Management and the Advisory Council for Housing, Spatial Planning and the Environment,* 2003.
http://international.vrom.nl/docs/internationaal/030601%20Summary%20RadenadviesEVH_engels.pdf

VROM, 2004a: *Besluit Externe Veiligheid Inrichtingen (External Safety (Establishments) Decree),* Staatscourant Sept. 23, 2004, nt. 183, 2004.

VROM, 2004b: *Guidance on the Accountability for Societal Risk,* The Hague, 2004.
http://www2.vrom.nl/docs/internationaal/Accountabilitity.pdf

VROM, 2005: *Safety In The Netherlands Five Years After "Enschede", 2005.*
http://international.vrom.nl/docs/internationaal/06%20Safety%20in%20t
he%20Netherlands.%20Five%20years%20after%20the%20Enschede%2
0fireworks%20disaster.pdf

VROM, 2007: *Guidance on Justification of Societal Risk,*
http://www.vrom.nl/get.asp?file=docs/publicaties/8043.pdf&dn=8043&b
=vrom

Western Australia, 1987: *Risks and hazards of industrial developments on
residential areas in Western Australia,* EPA Bulletin 278, Environmental
Protection Authority, Western Australia, Perth, Australia, 1987.
http://www.epa.wa.gov.au/docs/242_B278.pdf

Western Australia, 1990: *Review of the guidelines for risk assessment in
Western Australia - Information to assess public input to the
Environmental Protection Authority,* Environmental Protection
Authority, Western Australia, Perth, Australia, 1990.

Western Australia, 1992a: *Criteria for assessment of risk from industry,*
Bulletin 611, Environmental Protection Authority, Western Australia,
Perth, Australia, February 1992.
http://www.epa.wa.gov.au/docs/675_B611.pdf

Western Australia, 1992b: *Criteria for assessment of risk from industry -
expanded discussion,* Bulletin 627, Environmental Protection Authority,
Western Australia, Perth, Australia, May 1992.
http://www.epa.wa.gov.au/docs/691_B627.pdf

Western Australia, 1994: *Risk Criteria - on site risk generation for sensitive
developments,* Bulletin 730, Environmental Protection Authority,
Western Australia, Perth, Australia, January 1994.
www.epa.wa.gov.au/docs/1414_B730.pdf

Western Australia, 2000: *Guidance for Risk Assessment and Management:
Offsite individual risk from hazardous Industrial Plant,* Environmental
Protection Authority, Western Australia, Perth, Australia, 2000.

Western Australia, 2005: *Separation Distances between Industrial and
Sensitive Land Uses,* Environmental Protection Authority, Western
Australia, Perth, Australia, 2005

Western Australia, 2007a: *Dangerous Goods Safety (Major Hazard Facilities) Regulations 2007*, Western Australia, Perth, Australia, 2007. http://www.austlii.edu.au/au/legis/wa/consol_reg/dgshfr2007537/

Western Australia, 2007b: *Discussion Paper On Risk Acceptance Criteria For Major Hazard Facilities In Western Australia*, Department of Consumer and Employment Protection, Western Australia, Perth, Australia, 2007.

INDEX

A

acute exposure, 154, 156
acute health effect, 7, 82, 107, 162, 167
Advisory Committee on Dangerous Substances (ACDS), 55, 94
Advisory Committee on Major Hazards (ACMH), 52, 130
advisory goals, 75-76
aggregate risk, 93
American Petroleum Institute (API), 35, 39
anchor point, 53, 55-56, 58-59, 63, 88, 114, 129, 139, 142, 160, 181
apportionment, 92-94, 98, 106
as low as reasonable achievable (ALARA), 63-64, 69, 70, 72-73, 76, 162, 167
as low as reasonable practiable (ALARP), 37, 44-47, 50, 54, 57, 59, 67, 70, 73, 75, 76, 78, 85, 87, 127, 130, 134-135, 138-143, 150-151, 161, 175-176, 189-190
Australian Capital Territory, 143
Austria, 161
average individual risk, 32, 34, 101, 107
average person, 88
average rate of death, 36, 156
average unprotected person, 76

B

Bell Telephone Laboratories, 12
Brazil, 53, 130-133, 162-165
broadly acceptable, 44-45, 47, 55, 57, 59, 63, 67-69, 71, 84-85, 121, 127, 129, 131-132, 134, 137, 142-143, 150, 159-160, 162, 167, 181
BRZO, 30, 63, 65
Buncefield, 59, 124

C

Canada, 143, 164
Canadian Society for Chemical Engineering., 143
Canvey Island, 13, 56-59, 90, 160, 175
Center for Chemical Process Safety (CCPS), 2, 20, 106, 171
Challenger, 12
checklist analysis, 15
comparison of risk criteria in the UK and the Netherlands, 67
consensus standards, 38-39
consequence modeling, 14, 21, 26, 184
consequences, 2, 3, 5-7, 14-15, 17-18, 24, 26, 38, 41, 52-53, 59, 61, 82, 97, 106, 109-110, 145-147

Printed and bound by CPI Group (UK) Ltd, Croydon, CR0 4YY

17/04/2025

14658861-0001